Egbert Pomeroy. [from old catalog] Watson

How to run engines and boilers

With a new section on water-tube boilers

Egbert Pomeroy. [from old catalog] Watson

How to run engines and boilers
With a new section on water-tube boilers

ISBN/EAN: 9783743467002

Manufactured in Europe, USA, Canada, Australia, Japa

Cover: Foto ©berggeist007 / pixelio.de

Manufactured and distributed by brebook publishing software (www.brebook.com)

Egbert Pomeroy. [from old catalog] Watson

How to run engines and boilers

How to Run

Engines and Boilers

WITH A NEW SECTION ON

Water-Tube Boilers

PRACTICAL INSTRUCTION
FOR YOUNG ENGINEERS
AND STEAM USERS :: :: :: ::

BY

EGBERT POMEROY WATSON

AUTHOR OF: MODERN PRACTICE OF
AMERICAN MACHINISTS AND ENGINEERS—
MANUAL OF THE HAND LATHE—THE PRO-
FESSOR IN THE MACHINE SHOP, ETC., ETC.

FOURTH EDITION

NEW YORK:
SPON & CHAMBERLAIN,
12 CORTLANDT STREET

LONDON:
E. &. F. N. SPON, LTD.
125 STRAND

1899

PREFACE TO FOURTH EDITION.

THE author has carefully gone through this little work, and at the suggestion of the publishers has added twenty-eight pages of new matter treating upon the subject of Water-Tube Boilers, their management, maintenance, and efficiency for marine and land service. Eight cuts, illustrating the principal types of Water-Tube Boilers, have also been included. While the space given to the treatment of this subject is necessarily limited, still the publishers believe it will add additional value to the work and prove of considerable interest and use to the engineer.

THE PUBLISHERS.

New York, October 8, 1899.

CONTENTS.

PREFACES have gone out of fashion. They are usually excuses, or explanations, or apologies, or something akin, for having written what follows them. This little book needs no preface, but here is one and a dedication as well, for in this work I have endeavored to serve young American Engineers who have taken to the business seriously by mentioning a few troubles they are likely to encounter; but it is only mention, for the one thing which can not be imparted is experience. Time, study, and practice alone can give it. No man was ever made an engineer by a book or by rules, ciples and traditions of his business. The inexperienced will find a few of them herein.

This little book is dedicated to American Engineers, the men who have always helped me on my way and who have always kept faith with me; who have always held out the right hand of fellowship to me as man and boy, by their sincere friend and well wisher.

EGBERT POMEROY WATSON.

HOW TO RUN ENGINES AND BOILERS.

CHAPTER I.

The first thing to be done upon taking charge of an engine and boiler, new or old, is to examine the boiler thoroughly. No matter whether it has just come from the shop, or has been run for years, take off the man-hole plate and go inside yourself with a hand-lamp; after you are in look at all the water spaces, and see if they are clean, that is, without rubbish or dirt of any kind. Even new boilers are not free from this. There are various irresponsible persons about boiler shops who are not as careful as they should be, and it will be the exception rather than otherwise if you do not find a lot of things which are better out of the boiler than in it. In large boilers rivet kegs are often taken in to sit upon, and are not always taken out again; the staves are thrown down in the water space, from whence they will float out and get in the steam pipes by some mysterious happening. The water line is not so high as the steam pipe, by any means, but the staves get there somehow; so do bunches of waste, etc. Everything of any kind should be cleaned out of the boiler before water is run into it. To do this it will be necessary to take off the lower hand-hole plates, and with a small hooked rod dislodge everything that is loose in the boiler.

Spare no pains in this work, for it will be labor and money saved in future. When the boiler is thoroughly clean you can put the plates in again, but before doing this rub the gaskets thoroughly with plumbago, and they will not adhere to the plate where exposed to heat. This is a saving of time and gaskets in breaking joints in future. The old way of making joints upon hand-hole plates, with hemp gaskets slushed with white lead, has gone out of use, having been superseded by rubber gaskets made for the purpose. If you are remote from a city, however, and cannot get a rubber gasket, you can make a hemp gasket which will answer all purposes. Jute is the best material for this job if it can be had, but you are quite as likely to be out of jute as out of rubber gaskets, and if you have neither you can get a clothesline most anywhere. Unlay this and take the twist out of it; beat it with a flat stick so as to reduce it to its original fiber ; then braid it up again of the proper size for the job in hand. Find the right length for the gasket, and join the ends so as to form a ring of the proper size that will fit the plate ac-

curately : it should go on so tight that you
have to force it over the flange of the plate.
Cover this gasket thoroughly with white lead,
and then put it in its place. It will be absolutely tight if the work has been properly done.

CLEANING THE BOILER.

We have been assuming that the boiler you
have taken charge of is a new one, without
scale, but if it is an old one there is likely to be
a quantity of scale and dirt, which must be
taken out at once. The way to do this is to
us the best tools you can get hold of, or con-
trive for the purpose. The place to look for
dirt and deposits from the feed water is in the
bottom of the boiler furthest from the entrance
of the feed water, and in the parts that are the
coolest, if there are any, when the steam is on.
In a return tubular boiler this is generally in
the smoke-box end, which is not so hot as the
fire-box end; the quantity of rubbish which accumulates in a neglected boiler is astonishing,
and you must not be surprised if you have to
remove wheelbarrow loads. This dirt comes
from the solid matter in the water, both that
which is carried in mechanically from turbid
supplies, and that which is held in suspension
until set free by heat. Every gallon has a certain amount, and as many gallons are evaporated daily, unless removed weekly it soon

makes a wheelbarrow load. This dirt, by
backing up against the flue sheet, deprives the
ends of the tubes of water, which not only
steals part of the heating surface, but destroys
the ends of the tubes and flue-sheet by corrosion and over heating, so that it is only a question of time when the boiler will be practically
useless. If the lower course of the tube-ends
in the smokebox leak, be sure that they have
been abused in the manner stated. You will
probably find that they will leak after dirt has
been cleaned out. In that case the tubes must
be re-expanded, and to do this a boiler maker
must be called in. Do not, upon any consideration, try to tinker them up with a hammer yourself. You will only make a bad matter worse, and set other tubes to leaking which
were tight. Having taken out what may be
called the "loose dirt," though some of it is
very far from being loose, you will find
another job in front of you, and that is to get
out the dirt which is fast. In other words, the
scale. This is actual stone, artificially formed
within the boiler from the working of it. It
differs in character with the kind of water
used. If it is hard water, so-called, it will be
limestone scale ; if soft water, it will be sulphate of magnesia and soda scale ; either one
of them is bad enough, so far as the boiler is

concerned, and must be removed absolutely if there is to be any economy.

REMOVING SCALE.

There is a way to do this which we have practiced with success, and that is to run the boiler full of water up to the third gauge and then put in a quantity of a scale preventive. Of these there are numbers in market, but we do not name any one as the best. Doubtless none of them are wholly useless, though some of them are inert or do not act. You will have to find out by experience which one serves your purpose. Sometimes caustic potash answers a very good purpose. This when the scale is chiefly mud with sulphate of soda and magnesia combined. Caustic potash is the concentrated lye sold in grocery stores, but if wanted in large quantities should be purchased of wholesale druggists. To use it, dissolve it in a barrel of water, say 40 pounds to the barrel, and pour it into the boiler. This is about one-sixth of a pound of potash to the pound of water, and is strong enough for the purpose. After the purger is in the boiler build a light fire and heat the water to boiling point, and then haul the fire and let the contents stand. It is better to do this on Saturday night, if possible, leaving the water in the boiler until Monday morning ; you should then get up steam to, say, five or ten pounds pressure on the same water, haul the fire all out, and blow the boiler down. You will, in the majority of cases, find the boiler thoroughly clean, except for chunks of scale which cannot go through the blow-cock, and which must be taken out through the hand-holes.

Caution.

In handling caustic potash the utmost care must be used. It is truly caustic, or burning, and if a portion gets in the eyes it will cause serious trouble. The same is true of sores on the hands. Handle it with gloves; treat it very respectfully.

CHAPTER II.

If caustic potash cannot be had, a substitute may be found, in rural districts, in slippery elm bark. This is not at all caustic, but quite the reverse, being demulcent in character. How it acts we do not know ; but that it has a certain efficiency we do know, because we have cleaned boilers thoroughly with it. It makes little difference how much is used, put it in the boiler and let it stay there for a week, and there will be a benefit from its use.

These purgers just named are only of service where the scale is soft ; for hard scale a different one must be used, and to attack lime scale it should be of an acid character, for lime is alkaline, and its antidote is an acid. But just here trouble is likely to ensue in the hands of an inexperienced person. A good many will exclaim loudly against using an acid purger in a boiler, arguing that it will destroy the boiler as well, and that very soon. Some have shown us pieces of iron, that they immersed in certain boiler purgers, that were badly corroded. This is very likely, but it so happens that no boiler purger is used in that way. The purger is largely diluted with water, and acts very slowly upon the iron. It attacks the scale first, because it has the greatest affinity, or liking, for it ; after that it goes for the boiler plates; but there are no after effects of this character from a boiler purger, because it is no longer in the boiler when the scale is removed, and if a boiler is thoroughly washed out there is no danger to it from the use of a strong purge. There is very great danger from the presence of heavy lime stone scale, and since nothing but a purger with an acid reaction will remove it, we do not fear to use it ourselves. Some engineers have shown us boilers from which the scale was removed which had the plates badly corroded. This action was attributed to the use of the purge, when it was, in fact, caused by the scale itself. The corrosion was going on all the time underneath the scale, and when it was removed the injury it caused was manifest. Of two evils we are taught to choose the least, and in this case the use of a strong boiler purger is less than the injury and loss of fuel caused by scale. Get that out first, thoroughly clean the boiler after of all traces of the purge, and there will be no trouble arising from its use. It is understood, of course, that after the use of any boiler purge that the hand-hole plates must be taken off and the boiler cleaned out by hand, washed with a hose, and then filled up and blown out again before steam is raised. There is no middle ground or half-way measures possible in dealing with a dirty steam boiler. Get down to the naked iron and keep it so, inside and out, and the boiler twenty years old will steam as freely as one just out of the shop.

OIL IN BOILERS.

Do not upon any account put crude oil or any other kind of grease in a steam boiler. It generally gets in fast enough through the feed water where open heaters are used, without putting it in. The effect of putting oil in is, in a great many cases, to cause the crown sheet to come down, or the lower sheets to bag. When first put in the oil floats, but it gradually picks up scum from the surface, in which scum there is always more or less actual mud thrown up from the bottom by the boiling water; the oil then becomes like tar, and being heavy settles on the plates and sticks fast. Since the water cannot get underneath it the plates are overheated and come down, notwithstanding the fact that there is plenty of water in the boiler. Keep every kind of grease out of a steam boiler, if you have to filter the feed water to do it.

BRACES AND STAYS.

We have now a clean boiler to deal with; let us see in what condition it is as regards strength. The braces are the first to be considered. Perhaps some of them are carried away entirely; such a state of things is by no means unknown. They must be replaced at once by boiler makers, who should also go over every other part of the boiler and test it for condition; but if there are no boiler makers handy the engineer must do it himself. The boiler most generally used is the return tubular, which is a plain cylinder with an external fire-box, from which the heat traverses the bottom and enters the tubes at the back, passing through them to the breeching and smoke-stack in front. The weak point in the return tubular boiler is directly over the bridge-wall, where the heat deflected from the wall strikes upon the shell. This spot needs to be carefully examined, for unless the boiler has been well taken care of it wll be found weak and unsafe. If any doubt exists, a half inch hole should be drilled in the bottom to ascertain the exact thickness of the plate, when, if thinner than the shell elsewhere, it should be removed, and a new plate put in. The sides of the boiler should also be examined near the wall, or where the boiler meets the brick-work, for here there is often trouble from corrosion ; also at the junction of the blow-pipe in the bottom or at the end of the boiler.

CHAPTER III.

Mud Drums and Feed-Pipe.

If there is a mud-drum attached to the boiler examine it very thoroughly, for explosions of mud-drums are very common. If the mud-drum is buried, as it often is, it is probably corroded greatly. The proper place for a mud-drum is outside of the boiler, in plain sight, where it can be got at and cleaned out weekly. The object of it is to catch all the free mud, so to call it, which is thrown down at night when the boiler is not running. With some water used for steam making, as on Western rivers, the quantity of mud so deposited is very large, and if not removed it will be driven back into the boiler. This is particularly true where the feed pipe enters through one end of the mud-drum. It does not require much thought to see that this wholly defeats the object of the mud-drum, for the sediment which collects over night is forced through the boiler again at the first stroke of the pump in the morning.

As to the best place for the feed pipe to enter the boiler there is a difference of opinion among engineers, but there is no doubt but that the worst place is through the mud-drum, for the reason given. Some think that the feed should enter the coolest part; some put it in the steam space, and some enter it at the front end alongside the fire. An objection to this is the bad effect of water cooler than the water in the boiler upon hot plates; an advantage is the propulsion of any sediment that may lie upon the bottom of the boiler to the back end of it, where there is no trouble in removing it; another benefit is that the mechanical action of the entering jet assists the circulation by forcibly driving the heated water from the front to the back, and replacing it with cooler water, but to effect these objects the feed pipe must project within the boiler for a few inches so as to give what we shall call a straight shot from it.

Boiler Fittings.

In these are included every sort of attachment to a boiler, the water gauge, gauge cocks, safety valve, checks of all kinds, and the blow-off valve or cock for blowing the water out of the boiler. This last is a much more important detail than it is generally supposed to be, and many accidents have happened from careless-ness with it. These accidents occurred from "sticking" any sort of a bent pipe (we have actually seen an old leader pipe from a house used) over the end of the nipple or elbow on the blow-pipe. The blow-pipe connection should be made as firmly and as securely as any other attachment to a boiler. If one reflects for a moment, it is easy to see that there is a tremendous strain on a pipe which is discharging a two inch stream of water under 60 or 70 pounds pressure. A boiler never should be blown off at this pressure, but it sometimes has to be, and preparation should be made for it. No elbows should be used where it is possible to avoid them; the pipe should run as straight as it can from the boiler to the outer air, and if a cock is used care should be taken that it is always in perfect order. It should not leak a drop; the bolt at bottom which keeps the plug in the cock should be accurately fitted, of full length, entering the plug not less than 1½ inches, and have a good head on it. It must never be meddled with or touched, except to open or close it, when under pressure. Don't hit it with a hammer, either on the under side to start the plug up if it sticks, or on the top for any purpose. Remember that it is under pressure, and if it gives way it is almost certain death to any one near it. Defer all tinkering and investigation until the boiler is cold; or, in other words, make everything secure before steam is on; then there will be no trouble.

Forethought, care, and caution, are absolutely indispensable qualifications in an engineer, and it is useless to expect success or ordinary economy without them. No man can be an engineer worthy of the name who is careless or has not his wits about him at all times. Mr. I-Didn't-Think has no business with a steam boiler.

What has been said of the blow-cock is true of all the other fittings. Every one of them must be in perfect order to be safe and efficient; an engineer must bear in mind that he is dealing with a tremendous agent, which is safe only when in its place and under control, and every gauge or fitting of every kind must be securely in place and tight of itself, that is, tight without makeshifts of any kind.

The safety valve in particular must be tight, for a great deal of coal can be lost by a leaky valve. It should be free and clear in the hoist, or where the lever is to be raised—if it is of the lever type—and entirely free from rust in all parts. A safety valve is for use, not for emergency, and if it is not in order it will not act when the emergency comes, if it ever does.

It is not every engineer that can do the work which we have mentioned with his own hands, for not all persons in charge of engines are machinists; these instructions convey a knowledge of what is needed, and the work can be supplied by those competent to perform it.

CHAPTER IV.

GRATE BARS AND TUBES.

One of the most important parts of a boiler is the grate. Curiously enough but few give this matter sufficient thought, but it is plain upon reflection that the air which is needed to support combustion must be supplied through it. If the bars are warped and broken, too much air goes through them, with the effect of wasting fuel or checking the free steaming of the boiler. Moreover, a broken grate bar prevents proper firing, or attention to the fire ; all the bars should be,in good order, with no open spaces at the ends (front or back) or sides. As this bears directly upon the subject of combustion, it will be more explicitly alluded to further along in this work.

The tubes or flues particularly demand attention, and must be absolutely clean inside and out. In a former article we have given directions how to clean them outside—that is, on the water side, but they must be clean on the fire side too. With anthracite coal this is not a matter of difficulty, but with soft coal it is ; not so much through the soot which accumulates in them as with the " gurry," for want of a better name, which is burned on. This last is the tarry distillates of the coal, or heavier products of combustion, which are condensed on the inside of the tubes when the boiler is comparatively cold, or in getting up steam every morning, and is by no means easy to remove. It not only checks steam making by obstructing the heat from passing through the tubes, but it hinders the draught by the adherence of soot and roughening the surface of the tubes. It would seem that the fire should burn this deposit off, but it requires a much higher temperature to do this than that in the tubes, and the only way to remove it when it has accumulated in quantity is to thoroughly slush the tubes with crude petroleum oil, applied with a swab and allowed to remain for a day or so, when it should be swabbed out again. This is a job which but few persons care to undertake, particularly if the boiler is large, but in some cases it becomes neccessary. Crude petroleum is a solvent for tar, and will clean the tubes thoroughly.

Of course it has to be undertaken in holiday time, when the boiler is idle for a day or two, for to be of any service the oil must remain in the tubes at least 24 hours. It is of no use to try to rasp this "gurry" out with steel brushes or scrapers. It is as tough as India-rubber and a scraper slides over it.

BRIDGE WALLS.

The bridge wall in a boiler is intended to delay the products of combustion in the fire-box as long as possible, and to confine the heat from the fire within the area of the grate. To do this it is manifest that the throat, or opening over the bridge wall, between the top of it and the boiler, should be as small as it can be, and leave room enough for a good "draught," so-called. There is, however, a danger in this, and this danger is that if the throat is too narrow, the heat, and sometimes the flame, is sharply deflected and concentrated directly upon one spot over the wall. The result of this is that the sheet for a foot or so is fire-eaten, or thinned and weakened; it is burned, as boiler makers would say, notwith-standing there may have been plenty of water in the boiler. The opening over the bridge wall should never exceed ten inches, nor be less than eight inches, and it should follow the curve of the boiler. There are a great many patents on bridge walls which are in-tended to improve the combustion by admit-ting air over them, or through them, but never having had any experience with them we can-not say anything about them.

Assuming that the boiler has been put in good condition, we will look at the engine.

The hints given in the previous chapters
should enable any intelligent man who is fit
to be about a steam plant, to have a boiler
which will steam freely and as economically
as its construction will allow. A treatise
could be written upon boilers alone, and many
such works are in existence. The contents,
however, relate more particularly to the construction, a matter which does not enter into
an engineer's duties.

THE SLIDE-VALVE THROTTLING ENGINE.

The commonest form of steam engine in use
to-day is the slide-valve throttling engine,
which is regulated by governors of various
kinds. It is the simplest of machines, easily
managed by any one after a little instruction,
and frequently is found in charge of men and
boys who have had no experience whatever,
they merely knowing that a certain valve has
to be opened, and that the engine must be at
half-stroke to start. Such persons are not engineers in any sense of the word, for they do
not intend to follow the business any longer
than they can help. Our instructions are not
directed to them, but to intelligent young men
who have started with the intention of learning
all that they can. The first thing to do then in
taking charge of an engine is to see in what
condition it has been handed over, in order that

you may not be blamed for the sins of those who preceded you. The cylinder is the seat of power, and we want to examine it as soon as we can get a chance. If we have been under steam the day before, we leave the engine on the back center at night (Saturday night for instance), and take off the cylinder head. The piston is then at the end of its stroke, and we have an opportunity to see what the clearance is between the piston and the cylinder head. The latter detail will leave its mark on the cylinder after it is taken out, so it will be easy to measure directly from the piston to the said mark. Some clearance is necssary for safe working, but it should be just as little as possible; clearance is waste room that has to be filled with live steam at every stroke before any work is done on the piston.

As a rule excessive clearance is given in small engines, for no reason whatever, except that some builders appear to think that there is less danger of breaking down. Suppose that the cylinder is 12 inches diameter: then the piston should run within one-quarter of an inch of the head. If the piston is of that class where the follower bolts stick out the depths of their heads, it cannot run so close as this, and probably there is an inch or more clearance in such a cylinder, but it is easy to reduce the clearance in such cases to the lowest point, and this is easily done by taking the follower to a machine shop and having the bolt holes counterbored, so as to let the heads in as far as possible; having done this, fill up on the head itself by bolting on a cast-iron plate of the required thickness, cutting out where it covers the steam port. The reduction of clearance often makes a boiler much larger; or, in plainer terms, since less waste occurs it is easier to keep steam on a boiler than when the clearance is excessive. Having found what the clearance is on the back end then, we disconnect the piston from the cross-head, and (running the crank on the forward center), we find what it is on the front end. This we do by shoving the piston clear up against the forward head. Having done this we measure from the follower back to the end of the stroke, as shown by the wear on the guides, and the wear on the cylinder itself. If this measurement is half an inch longer than the working stroke of the piston, there is half an inch clearance on the front end, and as there are no follower bolts on that end it is all waste, except so much as is actually needed for the safe working of the engine. We should, if the engine was ours, reduce this clearance also, to the same degree that we did the back end, but as it entails more or less work for the shop, it will be as well to leave the clearance half an inch on the front end; if the clearance is one inch, however, no consideration of trouble or expense should be spared to reduce it in the same way that we fixed the back head, by adding to the head itself. The clearance in any engine must be reduced to its lowest terms, for by doing this, if the engine is yours, you put money in your pocket; if it belongs to some one else and you are in charge of it, you get the credit of making a saving, and this will be a feather in your cap worth working for.

CHAPTER V.

THE PISTON.

Now that we have the clearance matter attended to, let us see what kind of a looking thing we have for a piston. This detail of a steam engine is of all conceivable forms—and some inconceivable forms, to any one who thinks what a piston has to do. They are made as heavy as hydraulic plungers, and with as many attachments as possible, in the shape of rings, with springs to keep the rings out to the cylinder, and screws in the springs to keep the springs out to the rings. The reason that some firms make them in this way is because their grandfathers made them so, and that is reason enough in their eyes. If the piston you have taken out is of this class it is your and the owner's misfortune, but as we are not giving instructions upon how to build engines, we will merely state how this old-fashioned piston is to be put in as good condition as possible. Pistons are liable to become leaky in the following places : between their flanges where the rings bear; between the rings and the cylinder itself; through the follower into the body of the piston. Wherever there is a joint look for a leak, for joints become imperfect through use and time. If the rings move back and forth between the flanges of the piston they leak, and must be made tight by skinning off the follower. This is of course a shop job, with which the engineer has nothing to do, but before the piston is sent to the shop for repairs the engineer should be sure that the piston needs it. Very often it will be found by examination that dirt or "burrs" have got in between the follower and the spider, or else the thread on the bolt-holes in the spider has been raised around the edges, so that the follower will not go down, iron and iron. An experienced engineer will soon find out whether these things have happened by taking a smooth file and going carefully over the follower-seat on the spider, or main casting of the piston. In this way he will find all the burrs or bruises that have raised the surface, and dress them off level; then when he puts the follower on again and screws it up solid without the rings in, he should take a hammer and strike on the outside of the follower opposite solid iron. If the follower is tight on its seat it will sound like striking on an anvil; if it is leaky the sound given out will be like striking a piece of iron lying on an anvil. Leaks can also be told by the appearance of the parts, but as this is not easily conveyed in print we shall not attempt it. The best way

in all cases is to send the piston to a good
machine shop and have it put in perfect order,
and this is why it was taken out the first thing,
so that it might be going forward while we are
dismantling other parts of the engine.

THE SLIDE VALVE.

The next thing we do to ascertain the condition of our engine is to take the bonnet off the
steam chest and see in what shape the valve
and its seat are.   An inexperienced man is very
likely to get into trouble here, and do damage
to the engine.   Bolts and nuts which have been
long undisturbed are very hard to start, and in
very many cases they either break short off in
the casting, or else, in the case of stud bolts,
come away at the bottom, and unscrew from
the casting.   Either of these misfortunes is
bad, because it is not an easy task to get out a
broken stud bolt, or to make one tight in its
seat after it has been forcibly removed ; therefore, if the nuts do not yield to moderate force
exerted on a wrench, pour a little kerosene on
them and let them stand half an hour.   Kerosene is the most pervasive fluid known to the
trade, and it will seep into the most minute
crevices; if after its application the nuts will
not then start, get an iron ring, or a big nut
with some body of metal in it and heat it red
hot.   Put this over the stubborn nut until it has

become very warm and it will come away without any trouble.

If we digress here for a moment it is because the occasion seems to demand it. This digression is to again insist upon the necessity of care and caution in dealing with a steam engine. It is no evidence of skill for a man to go at a steam engine with a hammer and wrench and slaughter right and left, for by pursuing this course he can do more damage in a moment than he can repair in a day, and he can save both time and money by going at every job in a workmanlike manner.

The slide valve is really the heart of the steam engine, for upon its perfect condition and perfect action everything depends; if it is off its seat or badly set there can be no economy. When we take up a slide valve in an old engine we shall, in nine cases out of ten, find it in very bad condition. This is owing, in a great measure, to the way in which it is connected to the mechanism that operates it, and to the way in which it is constructed. Most slide valves are extremely faulty in this respect. In order to keep the steam chest as short as possible, the valve seat is made short, and very often the valve overruns the seat, so as not to wear a shoulder on it. The valve stem, acting on the stuffing-box as a fulcrum,

tends to pry the valve off its seat, notwithstanding the pressure upon it, with the result
that the face of the valve is worn rounding in
the direction of its stroke. Where this is the
case it must necessarily leak, for a slide valve
seat is like the slide valve itself—if one is
rounding the other must be hollow, in some degree, unless it is very much harder than the
valve itself. The time to test a valve for leaks
is when the engine is running, and it can be
told very quickly by watching the exhaust
where it can be seen. If this is sharp and
clear at every stroke the valve is tight, but if it
is followed by a secondary jet that scarcely
clears the exhaust pipe, the valve or the piston leaks, and quite likely both ; any leak
through the piston would also show on the exhaust, but in this case, unless the piston leaks
very badly indeed, it is likely to be a leak of
the valve which shows on the exhaust. To
test it for condition, obtain a straight edge and
lay it across. Hold the straight edge absolutely vertical, not tipped to one side, and it will
soon show in what condition the valve and the
seat are. The remedy for a leaky slide valve
is in the machine shop.

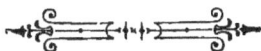

CHAPTER VI.

TESTING THE VALVE WITH RELATION TO THE PORTS.

To find out whether the valve is properly made in the first instance, or whether it has been tampered with by some engineer in charge before you, proceed as follows :—Take a sheet of paper large enough to entirely cover the valve seat and lay it on it. Rub all over the edges of the ports so as to obtain a fac

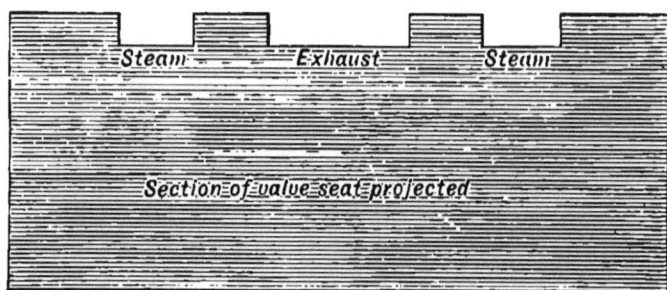

Steam   *Exhaust*   *Steam*

Section of valve seat projected

Fig. 1.

simile of them. Then get a piece of pine half an inch thick and three inches wide, and put the edge of it on the diagram, transferring the ports to the stick, thus :—Do the same to the valve, and you will have a fac simile of the valve and its ports, which can be more readily handled than by taking the valve itself, which is heavy and hard to see distinctly when in the chest. Now these directions sound very simple, and are very easy to understand by

one who knows all about the matter before-
hand, and who knows what he expects to see,
but they are not so simple to a young man
who reads them for the first time, or who is
unacquainted with the action of a slide valve,
and it is mainly to readers of this class that
this work is addressed. But we will try to
make it as simple as possible, and so that any-
one without previous knowledge of a slide
valve can see at a glance whether it is properly

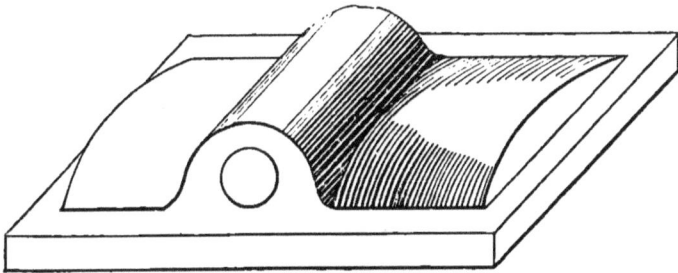

Fig. 2.

made or not. Actual comparison of the valve
and valve seat templets will appear further on.

Let us say, however, that there are slide
valves of many kinds, flat faced, round faced
(as in the case of a piston slide-valve), V faced,
etc.; but in this article, when we say slide
valve we refer especially to the common cast-
iron box without a bottom, which is generally
used in engines, as shown in the engraving,
fig. 2. This covers both ports and extends
some distance over them on each side. That

is to say, the end of the valve laps over the ports, and the part projecting is called the lap of the valve. The cavity inside the valve is the exhaust port of the valve, and this also laps over the exhaust edge of the steam port sometimes; the outside lap is called steam lap, or lap on the steam side, and the inside lap is called exhaust lap—when there is any. Usually the exhaust port in the valve coincides with the inside edges of the steam ports as shown in

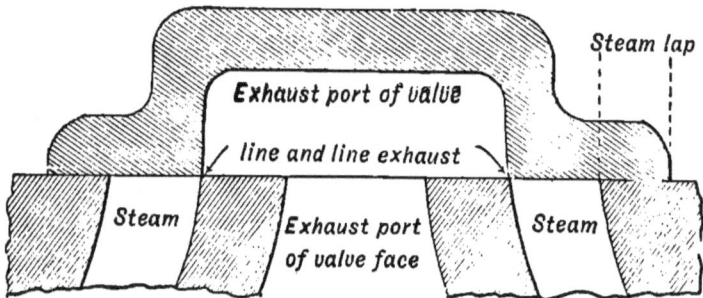

Fig. 3.

fig. 3, and when in this condition it is said to have line and line exhaust. Sometimes the exhaust is given clearance; that is to say, the steam port on the exhaust side is open slightly, and when in this condition it is said to have exhaust lead, or lead on the exhaust side. A slide valve then, works normally, that is to say naturally, under these conditions : It is a cast-iron box covering both ports all round, so that no steam can get into the cylinder unless the valve is moved so as to expose one of the ports. To recapitulate : the part which projects over the ports is called steam lap ; the inside cavity of the valve is the exhaust port ; the inside edge of the steam port is the exhaust side ; the outside end of the valve is the steam side ; and the same on both sides of course. These details are, naturally, familiar enough to experienced engineers, but we must not forget that there are young men coming into the trade continually who have all their trade before them, and who have it to learn as we had to, and it is for them that these explanations are given. Let us now look at the action of the valve.

Defects of the Slide Valve.

Were it not for one inherent, and we may say, hereditary defect, the slide valve would be the ideal one for its purpose, for all the functions are performed by one valve. This defect is that it is limited in its application to working steam expansively. As will be readily seen by any one who uses the templet, fig. 1, where the valve face is shown in section, when it is applied to the valve and moved to the various positions of opening and closing the valve, the exhaust is more or less throttled or choked ; its area is greatly reduced, so that escape of the exhaust is delayed. The result of this is that

the exhaust steam presses back on the piston
(back pressure so-called), and takes away just
so much from the power of the live steam on
the other side which is driving the piston forward. This back pressure varies in amount
with the position of the valve and the point of
the piston stroke at which the valve closes.
For instance, in plain words, when a slide
valve cuts off at three-quarters of the piston
stroke there should be little or no back pressure
in a properly constructed valve, for the exhaust
is open long enough to allow all the dead steam
to escape, but at points of the piston stroke
under three-quarters the exhaust is not free,
and a cut-off obtained with a common slide
valve under five-eighths of the piston stroke has
to be paid for by loss of live steam pressure.
Notwithstanding this fact there are many slide
valves cutting off to-day at one-half of the
stroke, and under that at times, and the de-
signers of them are satisfied—that is to say,
they have to be satisfied—for the common slide
valve will always create undue back pressure
at points under eleven-sixteenths of the piston
stroke. This is shown very plainly by indicator cards, where the last part of the exhaust
is caught in the cylinder by the piston and
pushed uphill, if we may so express it, until
(when nearly on the center) there is a pressure

opposed to the piston closely approximating boiler pressure. Whether this is economy or not every one must judge for themselves. To expend live steam pressure and power stored in the flywheel in trying to make dead steam alive, by squeezing it between the piston and cylinder head, always seemed to us unwise, for the reason that we do not get back as much work from the imprisoned steam as we spent to catch it, but as it is no part of our intention to discuss moot points or theories in this series, we go no further in this direction.

CHAPTER VII.

Lap and Lead.

The object of putting lap on a slide valve is to cut off the steam early in the stroke of the piston. Suppose the steam end of the valve had no lap at all, but barely covered the steam port: then so soon as the piston moved the valve would open and continue opening, closing barely in time to open again for the return stroke of the piston. Now suppose we add one-quarter of an inch lap to the valve; then the valve would open just as soon as it did before, because we have advanced the eccentric to permit it to open, but it would close sooner by the amount of the lap, because we have stolen, so to speak, a quarter of an inch from the travel of the valve by advancing the eccentric; therefore, if it closes sooner it cuts off the live steam earlier in the stroke; but, as explained previously, it cuts off the exhaust also. We introduce this as an illustration of the uses of lap. Laps on slide valves vary all the way from half an inch upon a twenty-five horse-power engine to one inch and upward on high power engines; on very large marine engines the lap amounts to 3″ sometimes; on locomotives it is usually one inch. If you have an engine which "takes steam all the way," that is, works full stroke, you can materially increase its economy, and to some extent its power, by adding lap to the valve upon the steam side ; the amount of it cannot be stated definitely, but must be governed by the size of the engine. Lead on a slide valve is the amount that the port is open to admit steam when the engine is on the dead center. The object of lead is two-fold : to have the ports and cylinder full of live steam the instant that the return stroke begins, and to check the momentum of the parts as they turn the center, or change the direction of motion. Now both the lap and the lead of a valve have an intimate relation to setting the valve for the distribution of steam, and as this will be alluded to further on in this series, we will say no more under these heads, because we shall be obliged to traverse the same ground, and this involves tiresome repetitions.

The Pressure on a Slide Valve.

Another defect or objection to a slide valve is the pressure upon it and the power required to drive it. This is great, though it is not so large as it is generally supposed to be. Specifically, in the case of small steam engines, Mr. C. Giddings, of Massillon, Ohio, made a dynamometer for the purpose of ascertaining the power required to move the valve on a

6⅜" × 10" horizontal engine. The surfaces were not given nor the pressures, but when exerting 13.5 horse-power at 200 revs. per minute, the power expended in working the valve was one-fifth of one horse-power. In an engine of 9" cyl. × 12" stroke, with a three-ported flat slide valve, at 100 revs. of engine per minute, the power required to drive the valve was 7.3 per cent. of the power developed by the engine, which last was 11.1 h. p. With a balanced slide valve on the same engine, at 100 revs., developing 15.6 h. p., the percentage of load on the valve stem was only 1 per cent. (*Mechanical Engineer*, page 62, vol. 12, 1886). This adduces an argument in favor of balanced valves *vs.* plain valves ; that is to say, the one is 6.1 per cent. lighter than the other to drive, but the fact remains that without any balancing but 7 per cent. of the power of the engine was required to drive it in a small engine. We do not say that this is not serious, nor do we think it unworthy of notice, but the fact remains that some valves require less pressure to work than others, owing to the manner in which they are lubricated and the condition of the seats. This last is the point we wish to make, for if the seat is cut the power required will be much greater than if it was in good order. Moreover, if the metal of the valve and seat are of the same degree of hardness, the valve will not work so well as when one is harder than the other. Of course the valve should be the softest, for it is easy to replace or re-face, while the seat is difficult to get at. The pressure on top of a slide valve is the steam in the chest bearing it down. When the engine is at work there is a pressure beneath the valve, reacting on the under side of its face, for the area of the port and through it. There is also a back pressure from the exhaust steam passing through the exhaust port of the valve; both of these pressures tend to reduce the direct pressure on the back of the valve, but to what extent can only be told by recording the facts in some particular case. The mean effective pressure shown by cards, as existing in the cylinder, is the pressure acting on the port-area face of the slide valve.

STEM CONNECTIONS TO THE VALVE.

We have said previously that one defect of the slide valve was its liability to wear untrue. One great cause of this is the manner in which the stem is connected to the valve itself. In locomotives the yoke is used exclusively. We believe there is not a single modern locomotive built without it, the reason being that there are no nuts or other details to work loose inside the chest.

This is of the greatest importance in an engine which is worked hard under high pressure constantly, but the yoke has its defects as well as all other mechanical devices. It frequently breaks, and at times cramps the valve so that it does not seat squarely ; it cannot be got out

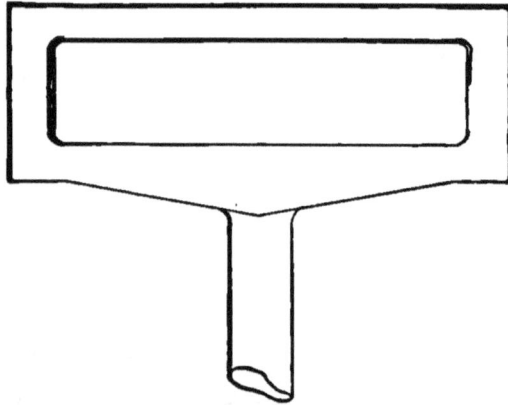

Fig. 4.

without lifting the steam chest, and it is also very heavy, and unless supported by the valve itself, wears away the gland very rapidly. Other common connections to valves are the nut in a pocket on the back, four nuts on a straight stem, the latter being run through a hole in the back of the valve, as shown in fig. 5 ; T heads on the stem are also common, the T fitting in a cross in the back of the valve. The nut in a pocket connection is one which is very liable to give trouble to engineers, for it is easy to see, unless the nut is exactly at right

angles to the travel of the valve, that it is apt
to cramp the valve and keep it off its seat. As
the stem is constantly wearing down the
trouble is of frequent occurrence, and it is difficult to detect when the engine is cold, for the
reason that the valve appears to be solid on its

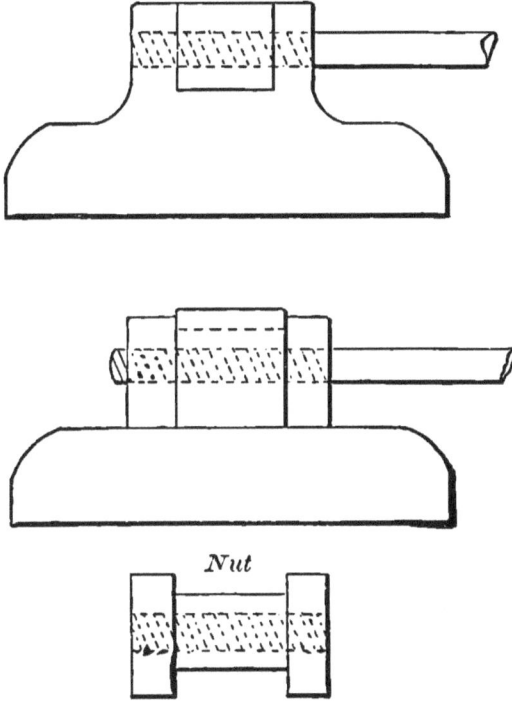

Nut

Figs. 5 and 6.

seat. We have seen engines which refused
work simply from this connection to the valve.
Upon opening the throttle the engine would get
steam under the valve and through both ports,
and nothing but easing the nut in the pocket
would let the valve down solid. Fig. 5 is the

nut and pocket connection, and the nut should in all cases be faced rounding on the working faces. A far better and simpler modification of this plan, and one we have used with success, is shown in fig. 6; it never fouls, and the nut allows the valve system to be lengthened or shortened without the use of jam nuts. It is easily put in or taken out, and fills all the requirements.

The solid nut arrangement shown is, to our way of thinking, the best. It holds firmly if properly fitted up, and it is also cheap to make, being all lathe work. It never cocks the valve or binds it any way; take it all in all, it is hard to find one better. These connections are the ones that are most commonly met with, and it is well to know what to expect of them.

CHAPTER VIII.

Valves Off Their Seats.

Now suppose we start or try to start our engine for the first time, and on opening the throttle find that the engine will not move, or will move as well one way as the other and without power in any direction. We know that steam is in the chest by the heat of it, and if everything was all right the engine should do its work; since it does not, there is plainly something wrong with the slide valve, and in nine cases out of ten it is off its seat. If it was simply wrongly set, the piston would go one way but not the other; it would make a great plunge forward or backward and stop there, but it would not drive the crank over the center. A slide valve does not require much to lift it from its seat, and it may occur at any time; a scale blown in from the steam pipe may get under one end and lift it enough to float the valve, then the steam will blow through the exhaust. When this is observed—blowing through—the remedy to be adopted is, in the small engines, to rap the valve stem smartly with a billet of wood, when, if the connection is in fault, it will frequently release the valve and allow it to seat itself. If something has got under the edge of the valve, move the stem as rapidly as possible back and forth, and it will work the obstruction off. If all these fail the only remedy is to open the chest and get at the valve itself. If water gets into the cylinder in any quantity it is very apt to jam the valve stem connection by bearing up on the under side of the valve through the steam port; it may even bend the stem in small engines. If this happens do not undertake any hammer and tongs remedies, but disconnect the stem, heat it black hot and straighten it with a mallet on a block of wood. Cold iron or steel breaks easily.

Valve Stem Guides.

In most modern slide valve engines the steam chest is on the side—right or left as occasion demands (usually the right), and the stem is directly connected to the eccentric rod without the intervention of a rock-shaft. The end of the stem is flattened, or squared, and is carried in a guide which may or may not be of service; if it is in line with the direct travel of the valve it is, but experience teaches that these apparently harmless guides can make a great deal of trouble for inexperienced persons, who fancy that the stem must move tightly in them. This is not so; the outer end of the valve stem must not be tied up in any way, but must be at perfect liberty, in order to allow the valve to lie flat on its seat. ' The only use of a guide on a valve stem is to prevent the weight of the eccentric rod from springing it downward, and to carry the weight of the valve stem itself; beyond this the valve re· quires no guiding, for the stem will attend to that. Do not, then, screw up the guide on the valve stem so tightly as to bind it in any way; it should work freely with a slight play in all directions.

GOVERNORS.

Let us leave the valve and all its connections, including the eccentric, until we get further in our investigations, and look at the governor or throttle valve. In early days engine builders made their own governors; these were always the common two-ball governors which regulated the engine (or pretended to) by means of a butterfly valve, so-called, in the steam pipe. This valve was merely a flat piece of brass with a shaft through it, hung in the steam pipe just as a damper is hung in a stove pipe, and usually one of these devices fitted about as well as the other. That is to say, the throttle was so badly fitted that it did not answer its purpose at all, and, added to this, its position in the steam pipe was such that it defeated its own object. The valve was so far from the steam chest that there was always a supply of steam between it and the main slide valve sufficient to run the engine at full power; consequently, when the load on the engine was reduced and the engine ran faster, the speed was not checked until the supply ran out, even though the governor had partly closed the throttle; then when the supply was worked off the engine slowed down, only to repeat the irregular motion at every change of load. Moreover, the old-fashioned two-ball governor was sluggish in its motions. The balls had to move through considerable arcs before the throttle acted at all; it had too many joints, which bound themselves tight by their motion, and it was so defective that it was cast aside for better devices. There are a good many descendants of the same family, however, still in the market, and they have the same inherent defects. The butterfly valve has wholly disappeared; at the present time no one makes them. Neither do engine builders make their own governors. Many patented governors for steam engines are manufactured by parties who make a specialty of them, and these makers use a simple cylindrical shell moving in a cylinder as a throttle valve. This works easily and tightly, and is a vast improvement on the old gear. Its faults are chiefly those of adjustment, and arise from neglect or carelessness on the part of those who run the engine. The parts are apt to wear, or else the stem gets lengthened by unscrewing, so that the valve drops from its natural position and blinds the ports. In caring for and repairing a governor all that is necessary is to see that the joints, when there are any (in some there are none, as in the Pickering), are free, the pins perfectly round and true, and free from burnt oil or gum; that the stem is straight, works freely and has no shoulders on it from working in one place constantly, and that the valve is in its proper place when the governor is geared up.

RUNNING WITH THE SUN.

There are a great many persons in existence yet who put faith in traditions, and who will gravely assure one that such or such a machine does not work properly because it does not "run with the sun." This is a notion that is firmly believed in by many who have faith, but no reasoning power. The sun has no influence upon, or any connection with machines made by man, with the sole exception of sun dials, and any machine which is in order will just run as well "against the sun" as "with the sun." Therefore, let no person impose upon you by telling you that the reason a bewitched governor does not work is because it runs against the sun. Suppose the engine stands east and west, how can it run against or with the sun? We used the expression "bewitched governor" in a figurative sense only, but let no engineer ever give up the search for a cause of bad working in a detail. It may be hidden, but it can be found by searching. There is always a cause for irregular action in all machines.

CHAPTER IX.

ECCENTRICS AND CONNECTIONS.

The office performed by an eccentric is to move the valve to admit steam at alternate ends of the cylinder. The eccentric is simply a wheel hung out of its own center. Its own center is a point equi-distant from the circumference. If hung on a shaft in this way it would have no other motion than a true rotary or concentric motion around the shaft, the same as a flywheel has on its shaft. Being hung out of its center, it has an untrue motion —an eccentric one—from which it takes its name. This explanation may sound somewhat puerile to experts, but there is an idea in the minds of many that an eccentric has some mysterious action which makes it especially fit for driving steam valves. We have been told by some that the eccentric ran fast and slow without reference to the speed of rotation of the engine, and it had, for that reason, a "dwell," so to call it, at each end of the stroke, that permitted the steam to enter quickly and to escape freely. The "dwell" exists, but it is is not by reason of any peculiarity of the eccentric itself, but on account of changing the motion of the valve from forward to back. At this period in the stroke the eccentric and all its connections are in line, see fig. 7, and for a portion of the stroke, from *a* to *b*, the eccentric exerts little or no effect upon its rod and the connections to it; in itself, however, it is moving at the same speed it always moves at,

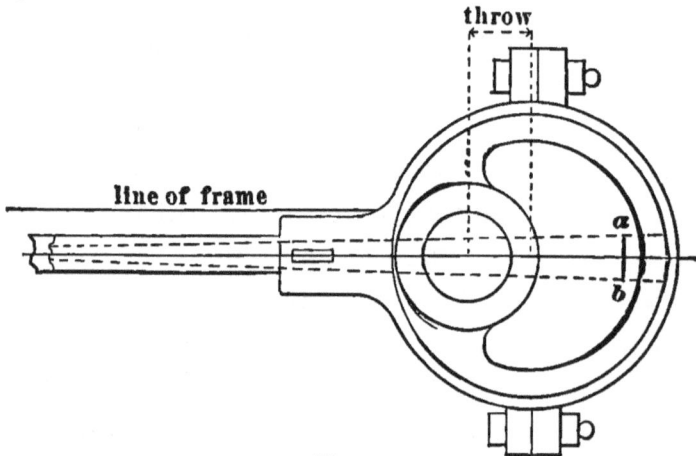

Fig. 7.

which speed is that of the engine. The idea that an eccentric has a variable speed doubtless arose from some one looking at the long side of it passing over the shaft rapidly, and comparing it with the short side, which does move slower than the long side, because it is nearer the center of the shaft. Now, an eccentric is hung out of its own center just half the stroke of the valve, because in a complete revolution it will double this throw, as it is called. The throw of an eccentric, then, is the amount it is out of truth (fig. 7), or the distance

from the center of the shaft to the center of the
eccentric. Suppose this to be 1½ inches,
then the eccentric is said to have 1½ inches
throw, and the travel of the valve is three
inches.

Connections from the valve stem to the eccentric are of various kinds. Where the steam
chest is on the side the eccentric rod is connected directly to the valve stem by a pin on
the side of the stem, or by a spade handle, as
it is called, worked on the stem itself. Sometimes, however, as when the steam chest is
not on the side, there is a rock shaft between
the eccentric and valve stem. This makes no
difference in the action of the eccentric, but
makes some difference in the position of it on
the shaft, as will appear later on in this series.
Sometimes there is an idler shaft, which also
rocks, but makes no difference in the position
of the eccentric on the shaft from that which it
occupies when directly connected. The connections are in all cases merely carriers or dis-
tributers of motion between the eccentric and
the valve itself, and need not be considered as
affecting the motion, except as hereafter appears.

THE CRANK PIN.

There is no more important adjunct of an
engine than the crank pin, for through it all the

power of the steam is transmitted. This statement does not refer to its office wholly, but to
its condition and its construction. In most
cases engineers are powerless to alter this without going to a great deal of expense, but they
can at all times keep it in good order, and in
such condition that the friction of it is reduced
as much as possible. Engineers worthy of the
name take the greatest pride in having this detail free from every scratch or flaw on its
working face, and, above all, never allow it to
get more than hand-warm; that is, about the
heat of the human hand. It should not heat at
all if properly attended to and when properly
proportioned in the first instance, but there are
many proprietors who run engines much beyond the power they were intended for, and
when this is the case the crank pin is liable to
suffer first. Crank pins heat from several
causes. When they have always run cool
with the normal load on the engine, and develop a tendency to heat when the load is increased, the cause is too much pressure per
square inch of surface; this forces out the oil
and brings the boxes into forcible contact with
the pin, so that heat is engendered. A remedy
in cases like this is to use a heavy oil, or a
grease composed of equal parts of plumbago
and tallow or lard. This finds its way into

the most minute ridges or imperfections in the bearing, and keeps the surfaces apart; it is a very excellent lubricant to use upon an over-loaded engine. More generally, however, crank pins heat from constant tinkering with the connecting rod end. An engineer hears a pound, and arguing at once that the crank pin brass must be slack, drives the key down, with the result of heating the pin. Now this matter of adjusting brasses on crank pins and on other bearings is an important one, not so well understood as it should be. In a great many cases the brasses are not properly fitted when they leave the shop, and are liable to cause trouble from that fact. High speed engines of the best class are properly made, for the builders of them are men of experience, but there are some persons who, as soon as they get charge of such engines, proceed to "relieve" the brasses in the wrong place, so that they can key them up. Now what is good for a high speed engine is good for a slow speed engine, and every bearing, no matter what its office, should bear "brass and brass," as the term is, and shown in the diagram at *a*—not as shown at *b*. The brasses should butt solidly and fairly together, and the pin should work easily inside of them. Then it will have merely the friction of work, and not the friction due to the work, with that due to the pressure of the key added. Many persons hold that no more pressure can be put upon a crank pin than that due to the work, and unless the pressure of the key or bolts exceeds that of the work, it adds nothing to the labor of the bearing. Those who hold this view are requested to try the experiment of driving in the key a little on a bearing which shows signs of heating. They will speedily

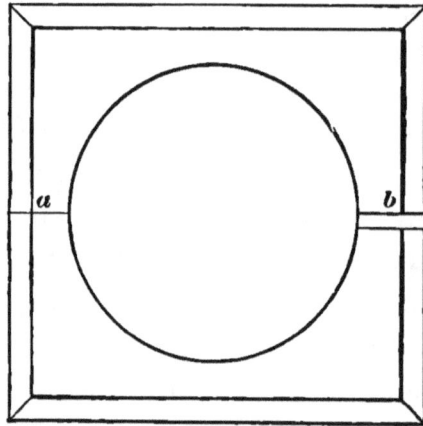

Fig. 8.

relinquish their theory. Another cause of heating of crank pins and other bearings is faulty workmanship. The brasses do not bear fairly or seat squarely and while they appear all right to the eye they are not all right to the bearing, which speedily gets warm over the matter. A crank pin brass must seat squarely on the end of the connecting rod, and the rod end itself must be square. If the key, when driven, forces the brass to one side or the other, and twists the strap on the rod so that its sharp edges can be felt on the side, it will draw the brass a-cock-bill on the pin, and make it bear the hardest on one side of it, reducing the area for working by the amount it is out of truth. The same condition of things is true of the main bearing. If the brasses do not bed fairly on the bottom of the pillow block casting, and do not go down evenly, without springing in any way, they will not run as they should. It matters not whether an engineer is a workman or not, in regard to his seeing these things. When they are pointed out to him, he can, and that is our reason for directing attention to them. If he knows where the fault is he can find men to remedy it.

Another cause of heating in bearings is too much surface in contact that is merely frictional. This is best explained by fig. 9, where all the work of transferring the power of the steam is done upon the surface of the pin, which is shown in section. All the bearing beyond this is of no service, but is a positive injury if if touches the pin, for it merely rubs and wears, without doing any good. Engineers then "clear" the brass on its sides as shown in fig. 9, for all bearings, whether those of the main shaft or elsewhere. We say "clear" the brass which means that it is to be just free, or so that it does not touch; not as shown in the diagram, where it has to be exaggerated to be seen at all. This clearance has another value, that of permitting the oil to stay on the pin, and to cover it at all times. This end is also furthered by cutting X grooves in the brasses, but this practice we have never been greatly in favor of, except in solid brasses which oscillate, or

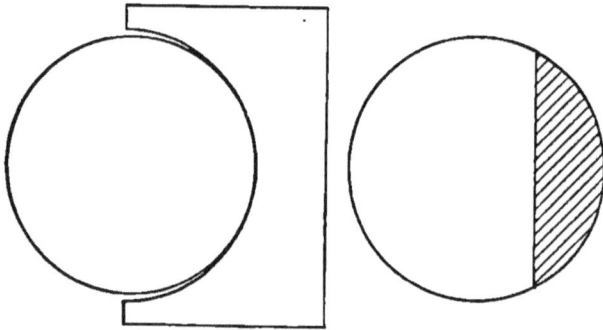

Fig. 9.

do not completely traverse the pin. For these last oil grooves are essential, inasmuch as when they are hard worked the oil is not distributed as it is in a complete revolution, and they are very liable to cut from want of access of the oil to all parts. Oil grooves, however, have the disadvantage of retaining dirt which may find its way in, they invite fracture, and they reduce the bearing surface. They are not to be used indiscriminately.

No greater annoyance can happen to an engineer than to have bearings heat beyond a certain degree. When shafts run hand warm it is no great matter, but it is better to have them quite cold, for then they do not give any anxiety lest they should become hot. Heat of any degree about a bearing is certain evidence of friction; what causes it is for an engineer to find out. If all bearings about an engine were absolutely parallel to each other, perfectly round, smooth, and true, of ample area and properly lubricated, they certainly would not give any trouble, but it is because some of the qualities above mentioned are lacking that they do give trouble. Want of proper materials in contact is also a cause of heating; dirty lubricating oil, or that which is too light in body for the work to be done, will also work badly for an engineer. Badly designed engine frames cause heating of main bearings by springing; settling of foundations, and badly fitted bearings do the same. For example, if on taking up a bearing that heats, the brass is found to bear as shown by the shaded lines in figure 10, the remedy is to scrape away the shaded portions so as to have a fair bearing, but before doing this an engineer should be sure that the fault is in the brass and not in some part of the pillow block, or other detail that holds the brass in its place. Brasses are usually made as light as possible to save material, and it is a very easy matter to spring them in fitting up. If they are so sprung it is of no use to refit the bearing itself, because that does not cure the trouble. It will continue to bear badly until worn out if the cause which springs it is in existence. Get the spring out first, and then refit the bearing, and there will be no trouble. Chronic heating in brasses is almost always caused by this defect—badly

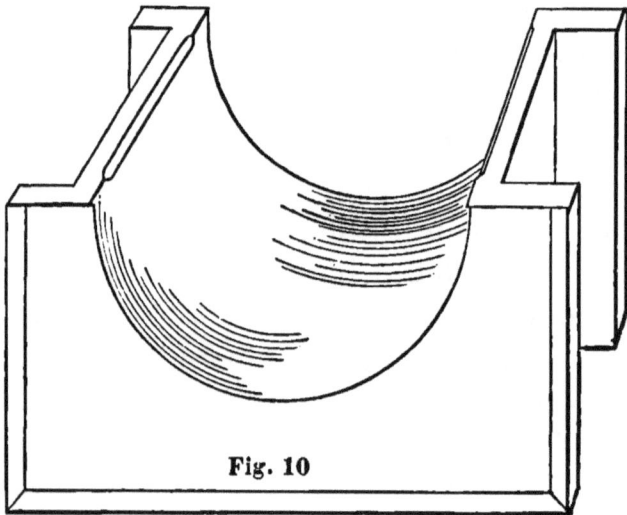

Fig. 10

fitting brasses. Another cause is, as stated, dirt, pure and simple. This need not be like sand or gravel to give trouble. Sometimes dirt gets in with the oil. All oil should be strained through a cloth, no matter how clear it looks. There is a great deal of dirt in lubricating oil of the average quality, as engineers find who strain it. Dirt also gets in through carelessness. Any work done on a floor over an engine shakes dirt down upon it at some time or other, and all floors over engines should be ceiled absolutely dust proof by laying paper between the planks. Imperfect lubrication is also a source of difficulty with bearings, though, as a rule, there is oftener too much oil used than too little.

CHAPTER X.

ADJUSTMENT OF BEARINGS.

Another, and perhaps a by far too common cause of trouble with bearings, is improper adjustment of them; that is to say, to the friction of the load proper is added the friction caused by excessive tightening of the bolts and nuts, or gibs and keys. It is easy to see, we think, that the office of a bearing is simply to hold the detail in its place while it is at work. A gib and key will not only do this, but it will also permit an engineer to take up a bearing as it wears, in other words, make it larger or smaller. Now, this is not a virtue, by any means, but a defect, for it gives an opportunity for careless men to do mischief through want of judgment. Men who do not think, so soon as they hear a pound or a noise about an engine, immediately accuse some bearing and go at it with a hammer or a wrench, and tighten it up. Bearings on an engine which is in line and in good order seldom require any attention of this kind. It is really surprising how long they will run without being touched in any way. We know of stationary engines doing heavy duty which have not had the crank-pin bearing touched in three years, and from which not a sound comes. It is the same with the main bearings; where everything is in good order they do not want any tinkering, and the best evidence an engine can give that it is not in order is noisy action. We know of some stationary engines that run at high speeds (240 revs. per minute constantly), yet no one would know they were running if they turned their back upon them. They are actually and absolutely noiseless. Not one penny has been spent upon them for repairs in over two years, and no tinkering of any kind has been done upon them. Facts like these prove our assertion that perfectly adjusted bearings and good workmanship combined will run satisfactorily for long periods. Unfortunately, not every engine is the best of its kind, and engineers can not always control the conditions. In other words they can not rebuild the engines, and we are willing to admit that there are some engines which it is very hard to "get the pound out of." Let it be borne in mind just what the office of a bearing is, however, and much can be done to lessen the annoyance of pounding. Reference will be made to this further on in this work, as some of it is due to faulty valve setting.

THE VALVE AND GEARING.

Having now gone from the cylinder head to the main shaft of our engine, and briefly re-

viewed the principal details, let us go back to
the steam chest again and look at the slide
valve and the valve seat, as shown in fig 1.
Let us compare them and see what relation
they bear to one another. The office of the
valve is to open and close the ports alternately,
as we all know, and if it is rightly made, it will
do this unfailingly, but it too often happens
that it is not rightly made, but is simply a cast-
iron box stuck in the steam chest anyhow, as
we may say. Sometimes, in small shops (and
in large ones for that matter), foremen get
notions in their heads that a slide valve was
never made until they got one up, and the man
who is afflicted with an engine of this kind has
a big bill for fuel. At other times engineers
themselves get notions as to exhaust lap and
exhaust lead, and cut away or add to slide
valves that were in perfect order before they
meddled with them. We have no theories of
any kind to propound, and no hobbies to ride,
and shall illustrate, therefore, only the usual
defects and the methods of curing them, leaving every one to adopt or reject them as they
see fit.

Figs. 11, 12, 13, show the slide valve in
various positions: the first one at mid-stroke,
where it covers both ports; the second with
lead, or just opening the port; and the third

with the port full open. This valve is shown as having line and line exhaust, that is to say, without lap on the exhaust side. The result is shown by looking at *a*, fig. 12, where the steam is passing out, as shown by the arrow; it has a free exit, to the extent of half the steam port

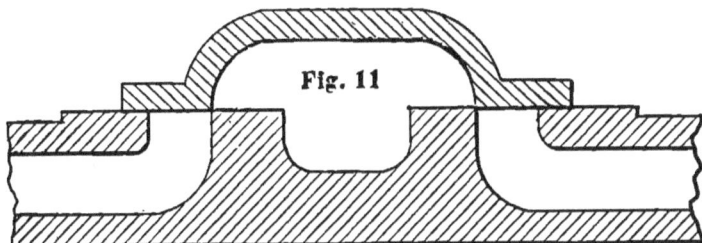

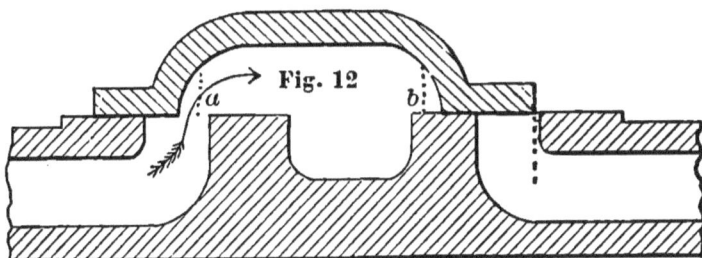

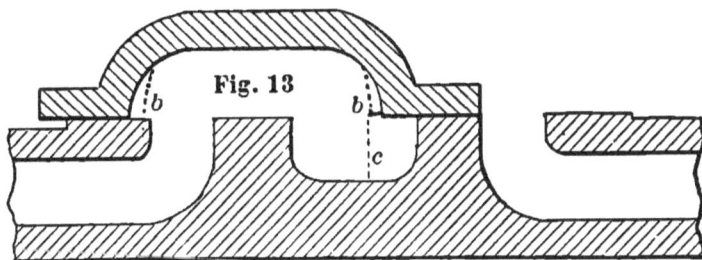

Fig. 11

Fig. 12

Fig. 13

nearly, when the crank is nearly on the center, but the exhaust began to open before the piston arrived at the end of its stroke. This is just the point where, it is claimed by those who are

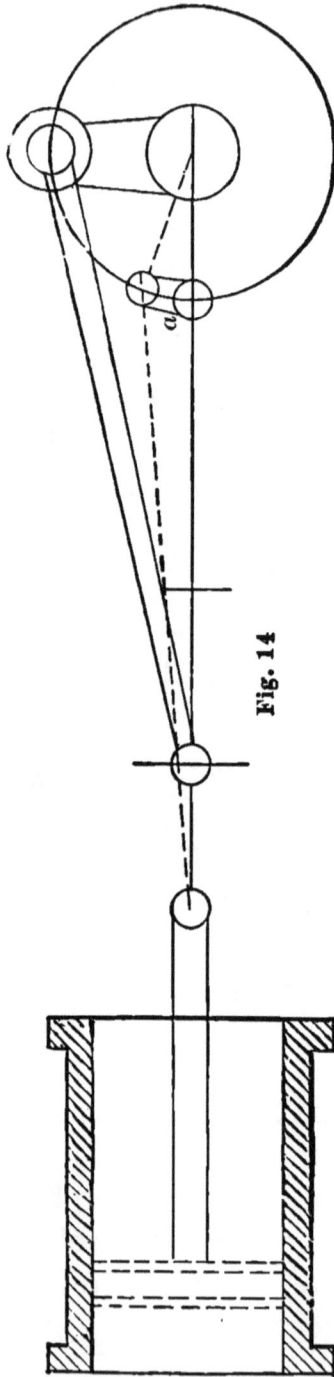

Fig. 14

in favor of inside lap on a slide valve, that an
error is made, because it lets the steam escape
before it has done all the work that it can.   In
some measure this is true, because every inch
that a piston travels under pressure gives
power, but the diagram, fig. 14, shows, to our
mind, that the steam on the last quarter of the
piston stroke does very little work indeed.   It
is at a comparatively low pressure, having
been expanded through the cylinder, and the
force exerted by it is spent upon a crank whose
radius is shown at *a*, fig. 14, and not in a direct
line, or at right angles with the line of motion,
but at a very obtuse angle, as shown by the
dotted lines, so that the effort to turn the crank
is absorbed to a great extent before it reaches
the shaft itself.   Suppose we do add inside lap,
as shown by the dotted lines at *b*, figs. 12, 13,
to the extent of half the steam lap, then we re-
tain the steam in the cylinder until the piston
has completed its stroke; we follow it up
with spent steam until it begins the return stroke; we get a full exhaust of
the spent steam through the steam port, but
we lose nearly half the area of the exhaust port
in the valve seat, so that, as shown at *c*, fig. 13,
which ever plan we adopt, whether line and
line exhaust, or lap on the exhaust side of the
valve, we have to sacrifice something, and the

general sentiment of experienced engineers is in favor of a line and line exhaust. The first thing an old engineer does who finds a valve with inside lap on it, is to chip the lap off, and swear some at the man who put it on.

In saying this, however, we must qualify it to this extent: that there may be cases where line and line exhaust is inadmissible or undesirable by reason of the proportions of valve face and steam ports. Our diagram shows the usual proportion of good practice. This is, that the bridge or metal between the two ports (steam and exhaust) is equal to the width of the steam port, and the exhaust is twice the width of the steam port. Not all valve faces and steam ports are so made, and the change involves some changes in valve construction and operation of the engine; but it is manifest that we can not treat upon this exhaustively in this work. The best arrangement for the exhaust must be determined for each engine by an indicator, which is the only friend an engineer has to tell him what is going on where he can not see directly.

These examples, it is understood, all exhibit the action of a slide valve when working at full stroke, or without cut-off, and from them it is easy to see that the evils of choking the exhaust and wiredrawing the live steam (that is, admitting it through a very narrow opening in the valve face), are increased when steam is used expansively, hence the unfitness of a slide valve for an automatic cut-off is readily understood. It is especially seen in locomotives,

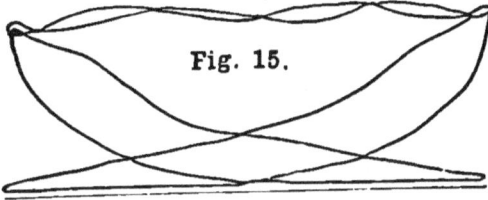

Fig. 15.

where, with a short cut-off, used at high speeds, the exhaust is actually punched out by the piston, for it begins to be compressed at half stroke, as shown by this card, fig. 15, which was taken from a locomotive on a fast run.

CHAPTER XI.

This detail is one of the simplest duties an engineer has to perform, but it is sometimes made a very mysterious matter. Elaborate preparations are made ; much peering into the steam chest takes place, and the chief performer looks very wise. There is no occasion for performances of this character, for the whole operation from first to last should not consume ten minutes, and a man of experience can set an eccentric at the first turn over, generally, after the valve is squared. This last means that the valve shall open both ports alike. Squaring the valve also makes the eccentric rod of the proper length, and until the valve is squared no setting of the eccentric can be done. Take notice that it is the eccentric which is to be set, not the valve. The valve occupies various positions on the valve seat, but the eccentric has a fixed position on the shaft for each particular valve. In one work on the slide valve the operation of setting an eccentric occupies two pages of directions, and endless *a b's*, *c d's*, *x y's*, and other letters of reference which are wholly useless. We never found any italics on valve stems, or on eccentrics ourselves. Moreover, much time is spent with trams, etc., in getting the exact mathematical center, or putting the crank pin exactly midway in its orbit, all of which is useless work. An eccentric can be set without any heavy flywheel to turn, or connections to drag hither and yon. Every part of the working gear not actually connected with the eccentric and valve should be taken off, for it only creates friction for nothing.

THE ACTUAL OPERATION.

Take out the crank pin (unless it is riveted in) and run a line through the cylinder.

Put on the eccentric strap and connect it to the valve stem, just as if it was under steam.

Now turn the crank shaft the way the engine is to run by any means that will turn it. If the flywheel is on use that for a lever.

Look in the steam chest and see if the valve opens both ports equally.

If it does not, shorten or lengthen the stem half the difference, until the eccentric moves the valve properly.

Now put the crank on its center by the line, and move the eccentric around on the shaft until it opens the port slightly, and stands as shown in the diagram, figure 16, at 1.

Turn crank 1 on the other center, and the valve will show more or less off of the position it had when on the other center. Divide the

difference by lengthening or shortening the
valve stem half the amount of error (for what

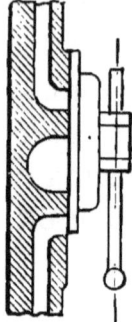

Fig. 16.

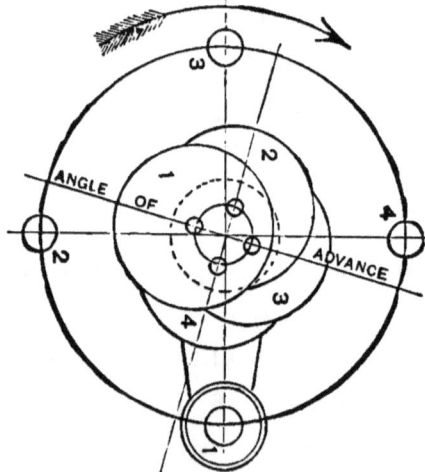

is taken off one end is put on the other in one
revolution), and the work is done.

There is no occasion to have the connecting rod or the piston in ; they have nothing whatever to do with setting the eccentric. This should be done the first thing after the shaft is in place, not when the details are all in. Once set, the eccentric is always set, unless it is shifted by chance. When it is once in place, it should be marked with a chisel, so that it can be put back if accidentally slipped

There are a good many who will object to this method of setting an eccentric, because it is out of the usual way; but it is as exact in result as any other way. There is no use in fussing with trams to get the exact mathematical center of the flywheel, because (unless the valve has no lap) no engine takes steam on the exact center. It always has more or less lead, the amount of which must be finally adjusted by an indicator for the work to be perfect. It will be seen by figure 16 that the eccentric is slightly in advance of the crank; that is to say, that its center line is not the crank's center line. The angle so formed is called the angle of advance, and the advance is made to take up the lap and to give lead, as shown in fig. 16. We must here say that in all cases the valve will be "late," as it is called, on the back end steam port, and this port will not open so fully as the front end port. As the explanation of

this involves a diagram, which, owing to the
limits of the page in the work, would be very in-
tricate, and not at all clear to inexperienced
persons, we shall not attempt one, but say that
the error is due to the fact that a fixed point on
· the eccentric rod in one revolution, and a fixed
point on the connecting rod in one revolution,
forms cycloids of diverse areas and outlines,
and a fixed point in one is not and never will
be coincident with a fixed point in the other;
the eccentric rod is always behind, varying in
degree with the length of the connecting rod.
If the latter was of infinite length, there would
be no difference in the action of a slide valve
on both ends of the cylinder, but the shorter
the connecting rod is the greater its angle of
divergence with the path of the eccentric rod,
and the greater the error in the valve motion.

Suppose we could drive a nail through the
side of a connecting rod, and could hold a
board up to it while the engine was running:
then the figure described would be a cycloid, or
egg-shaped. Now drive a nail through the eccentric rod, and the figure described by it
would be a cycloid also, but different in out-
line and area, and the shorter the connecting
rod the greater would be the discrepancies.
This is, in brief, the cause of the difference in
port opening for both ends of the cylinder, but

it affects the action not at all. There are many
valve motions which seek to overcome this so-
called evil, and such motions are called radial
valve gears, for they operate the valve by levers
instead of eccentrics ; some of them have eccentrics also ; one only for both motions, forward and back. These are used chiefly on locomotives and screw propellers, and the cycloid described by the valve stem connection
is a very close approximation to that of the
connecting rod, so that the port openings are
practically equal, and the cut-off is equal for
all points of the stroke. This makes a better
distribution of steam, and raises the efficiency
of the whole machine to some extent ; but the
actual values of these gears is very slight when
compared to the cost of keeping them up, and
their inaccessibility when in motion. The object of putting lead on a valve is to fill the
ports with live steam, for one thing, and to
check the motion of the piston gradually so
that it will cushion on live steam. The amount
of lead varies with the character of the work.
Engines which run slow, say 50 to 60 revs. per
minute, require very little, but high speed engines should have more. Say that our piston
is 12″ diameter and the engine makes 60 revs.
per minute; then 1-32d of an inch is ample lead
for the valve. A great deal of steam can get

through an opening of this dimension, but if
the same engine makes 400 revs. per minute,
then the valve should have 3-32ds lead, and may
even require more.   Now, the link motion, as
most persons know, increases the lead on the
valve as the steam is cut off.   It is useful to
bear this in mind, but we shall not attempt an
explanation of the cause.

This work, as its title indicates, gives elementary instruction upon the operation of engines and boilers, and we do not mean to go
outside of that and make it a medley of several
different branches of an engineer's profession.
Moreover, if we attempted this line of instruction, we should only repeat the researches of
others.   Those who want a treatise upon the
link motion and its operation should purchase
"Link and Valve Motions," by Auchincloss, which
is a standard work, explained in the clearest
manner, and fully illustrated.

Setting the eccentrics of a link motion is
precisely the same operation as that of any
other, with this difference only : that both
rods, back and go-ahead, must be adjusted for
length before the eccentrics are set, for a
change in one affects the action of the other.
Suppose, for example, that we undertake to set
the go-ahead side before the backing side is
corrected for length. We get the go-ahead

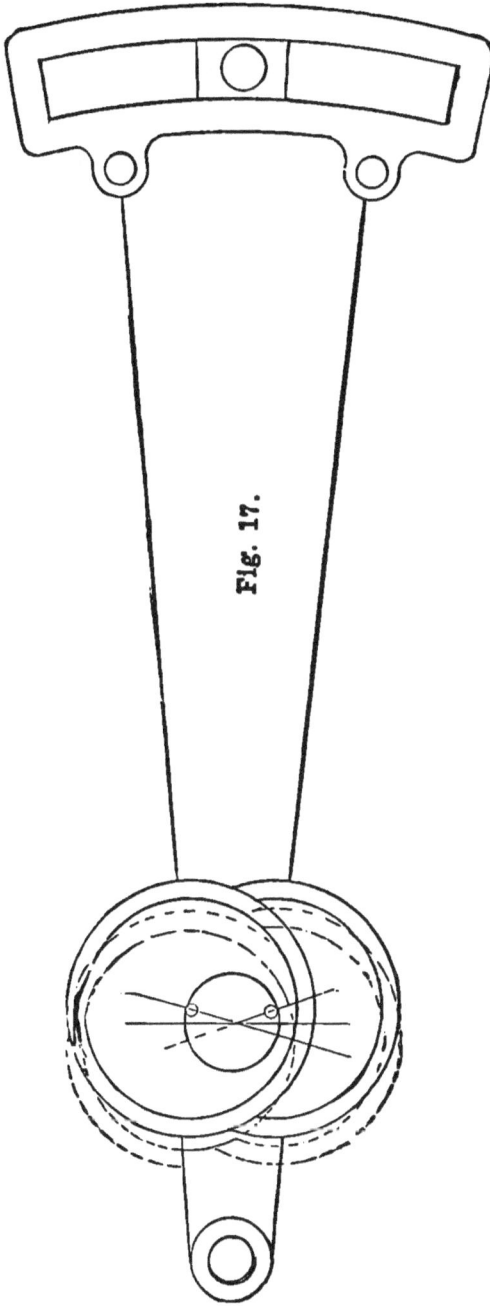

Fig. 17.

wheel on all right, but when we throw the backing side in gear and set it, then we find, on trying the go-ahead side, that it is out. This is caused by the backing eccentric rod being of the wrong length. Now, if we have squared the valve for both motions, we set the eccentrics as shown in this diagram, fig. 17. This position answers only where the link is directly attached to the valve stem. Where a rock shaft is used the dotted lines show the position of the eccentrics.

Setting the cut-off valves of an engine is a very short job. Usually these valves are so fitted that they operate from no admission at all—zero, so called—to full stroke. Sometimes, however, they are not connected to the governor, but are set to cut off at some fixed point, say one-half of the stroke. To do this, or to cut off at any point of the stroke, it is only necessary to square the valves in their travel over the main valve, run the crosshead to the point at which it is desired to cut off the steam, set the valves so that they just close the main steam valve port, and then turn the eccentric around on the shaft until it will connect with the cut-off valve gear. Or, connect the cut-off valve gear with the eccentric and then turn the same on the shaft until it just closes the ports at the desired point.

CHAPTER XII.

RETURN CRANK MOTION.

The return crank motion is the same as an eccentric, and is set in the same way. It has, however, the disadvantage that the lead can not be changed unless there is a slot for the pin *a* to move in, for, as will be seen in fig. 18, moving the return crank in or out from the center of the main shaft merely increases the travel of the valve, the lead being very slightly affected.

It must be borne in mind that setting the eccentrics of an engine is at best a haphazard operation when it is done while the engine is cold. Many changes take place in the valves and valve motion when the engine has been run a while and after it has been heated up. Cast-iron expands materially by heat, and the valve gearing itself stretches, if we may so term it. That is to say, the strain imposed upon the several joints and connections make the actual relation of the valve and eccentric different from what it appeared to the naked eye when the steam chest was open and the engine was cold.

This is one reason why it is unnecessary to waste time in finding "absolute centers" of en-

gines with a tram. All have to be corrected
by the indicator at last, for that is the only in-
strument we have for detecting the actual se-
quences of the valve and valve gearing generally. We venture to say that a good steam
distribution, as shown by the evidence of a
card, will bear very little relation to the absolute centers, or point of no motion of the
crank and piston.

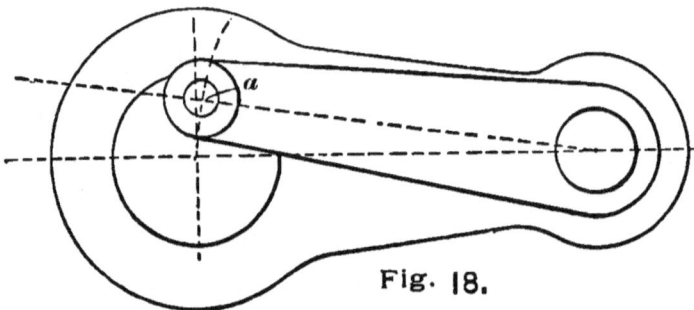

Fig. 18.

So far as setting eccentrics is concerned,
many of them, in radial valve gears particularly, are forged solid on the shaft. They are set
in the drawing room, and the relative lengths
of the several rods are plotted out on the drawing board. It must be borne in mind also that
this method of setting an eccentric, as to its
relative position with the crank pin, refers only
to the valves which slide on their seats, piston
valves included; not to puppet valves, or to all
forms of radial gears. The great majority,
however, stand as shown in the preceding dia-

grams. The influence the valve has on the action of an engine is very great. As we have said in previous lines, the principal object of it is to distribute the steam properly in the cylinder at each stroke, and incidentally, to correct or check the motion of the several parts at the end of the stroke. This is done by cushioning the piston on a bed of live or dead steam, as the fancy, or the teaching, or the experience, of the engineer directs. If all steam engines were perfectly built this would not be necessary, but there are many defects of construction and erection which have to be compensated for, and this is generally done by compressing the dead steam, or admitting live steam in the form of lead. The crank itself is one of the most perfect details ever devised by man for gradually absorbing or taking up the momentum of the parts, and many first-class engines are at work turning their centers with neither lead nor compression, and are also noiseless in action, but this can not be done with the average engine, or with very high speed engines, so lead and cushioning, or compression, is resorted to.

POUNDING.

One of the commonest defects of steam engines, and one that is the most annoying to hear, is pounding, so called. That is to say that in passing the centers a noise is heard, which may proceed from several causes, and is distinctive in character for each one. These can not be described so that an inexperienced person can tell the cause of it from the noise. Detecting or locating natural noises, so to call them, or the sound caused by the natural working of an engine, and separating them from noise caused by defective action, is a part of an engineer's duties which can only be gained by experience; so we shall not attempt it, but proceed to point out some which are common. The first we shall mention is when an engine is out of line.

Take a chalk line and stretch it taut, so that it is absolutely straight, without sag. This represents the center line of an engine. Now, if every detail is exactly in harmony with this, there can not be any sound from an engine. If the main shaft is exactly (mind what exactly means!) at right angles with the center line, and the path of the crank is exactly true also, if the crank pin is absolutely square with the center line of the shaft and revolves mathematically exact with it, then the engine is in line, and the fault is not there.

THE CONNECTIONS.

Now look at the connections. Suppose the connecting rod is not properly fitted up or

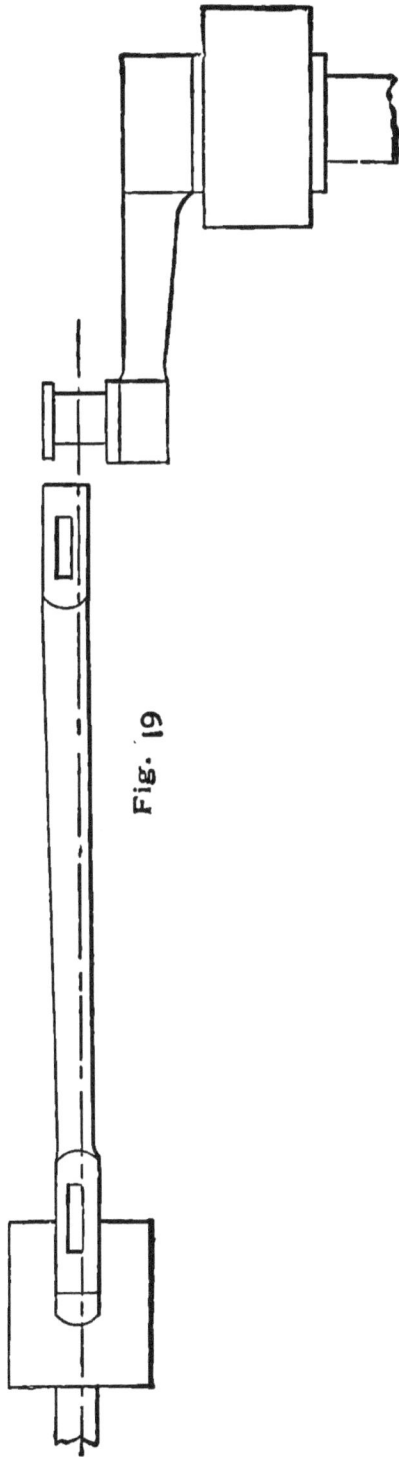

Fig. 19

keyed up, and stands off from the crank pin when cast loose from it, as shown in the diagram, then a noise will be heard which is caused by the "chugging" of the crosshead against the inside of the guides at each stroke; the crank pin springing the crosshead from side to side at each revolution. This noise is hard to locate when the engine is at work, particularly if the guides stand vertically and are V-shaped, as in Corliss engines. If it is suspected as a cause of pounding, disconnect the crank pin end of the connecting rod and key up the crosshead end tightly; then the rod will show for itself whether it hangs square or not. If it does not point fair for the exact center of the crank pin, between the collars, ease off on the back of the crosshead brass slightly, so as to throw the rod in the center of the collars. Try the crank on both centers, and if the rod shows off on each side, right and left alternately, then the main shaft is out of line and must be brought true. Where there is a heavy belt dragging on the outboard end of a main shaft it is very apt to haul it around materially, even canting the whole outboard foundation sometimes, if the latter is high and narrow, as it usually is.

We have stated before that it is seldom that bearings are so slack as to cause a pound. The noise of a slack bearing is a " chuck," so to call it, and not a pound proper, and it is of no use to endeavor to silence a noisy engine by screwing or keying up; that only makes a bad matter worse. Very often pounding is caused by improperly set valves, and sometimes a little more lead or less compression will cause it to work better.

This, it must be borne in mind, is only a rough and ready method of finding whether the crank shaft is true or not. The proper way to do this is to take the engine apart, run center lines through the cylinder and the guides, and see whether they are in line with each other. A plumb line should be dropped over the exact center of the crank shaft, in the crank pit, so that the vertical line barely touches the cylinder line. The crank pin should then be tried by this line, so as to ascertain whether it is equally distant from it on top and bottom centers.

Sometimes it will be seen, where the engine has run a long time, that the shaft needs to be raised at the crank end in order to bring it square. This is shown by the center in the shaft itself, by putting a square on the end of the shaft and allowing the blade to come gently down to the horizontal center line through the cylinder. These directions are very easily understood by persons who have had experience, but those who have not, need to exercise care in using the square, for if the end of the shaft itself is not true the indications of the square are of no value. Do not undertake to true any horizontal shaft by a spirit level. Shafts are not the same size all the way ; that is to say, they are not true themselves, notwithstanding that they may have been turned and are apparently true.

Another and very common source of pounding in an engine is changing the valve motion, or re-setting it, and overhauling the engine for repairs. If the valve time is changed so that it takes steam at a different point from where it formerly did, earlier or later, the pressure comes in a different place on the crank pin and shaft, which are worn to the old valve motion. The engine will not work then smoothly until the brasses are refitted. All bearings upon engines that have been overhauled are liable to this trouble, and it is an exceedingly difficult one to detect. The best way to avoid it is to re-bore all brasses and re-turn all bearings that can be so treated. The main shaft can not be, but the crank pin may be "skinned over," as it is called, to its great improvement. These things properly belong to refitting an engine in a machine shop, but as it is part of an engineer's duties to know them, we have said a few words in that direction.

A few lines back we made a brief reference to lining up an engine in order to avoid pounding, but perhaps a diagram and more explicit directions as to putting all parts in line will be acceptable. An engine out of line never will work as it should, and as it is a very simple matter to have it square we shall give plain directions how to make it so.

The diagram, fig. 20, shows a side elevation of a Corliss engine. In the crank-pit is a square frame made of boards, stiff enough to hold a line firmly without jar or tremor. This frame is not necessary if there are timbers overhead or at the end of the frame to fasten a line to; the frame is only put in the diagram to show the process. The piston, crosshead and connecting rod are taken off and out of the engine, and a line, *a*, is stretched through the cylinder. One end of it is fastened to the cross, *b*, shown in diagram *C*. This cross is made of wood firmly fastened together and having a hole in the exact center of it about ¼ of an inch in diameter, or larger than the line which goes through it, and the line is held by a piece of wire run through a loop in the end of the line, so that by moving the wire one way or the other, the line can be centered in the cylinder independent of the cross. This cylinder line **must** be as fine as possible, hard-twisted

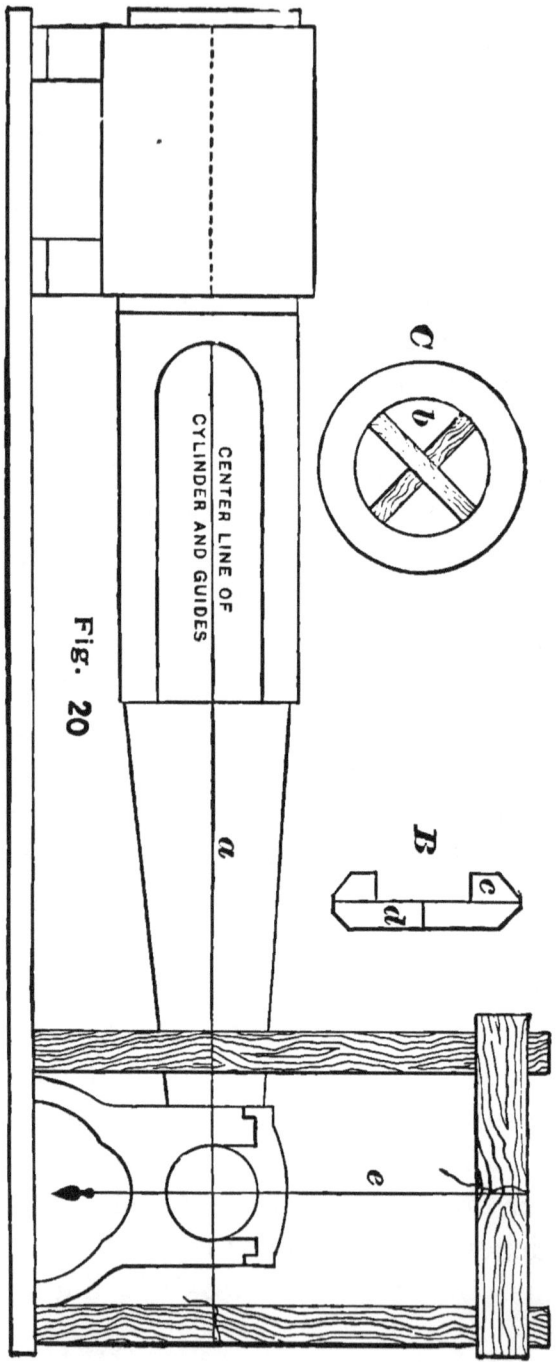

CENTER LINE OF CYLINDER AND GUIDES

Fig. 20

and very strong, so that it can be stretched very tight and have no sag whatever. Run this line through the cylinder, draw it up tightly, and then center it absolutely in the cylinder, by cutting sticks half the diameter of the cylinder, moving the crank end of the line until it is absolutely centered in the cylinder at both ends of it. Pay no attention to where the crank end of the line is ; let it go where it will with reference to the shaft itself. Now make a template, *B*, which just fits the guides accurately, and draw lines through it, as at *c* and *d*. Where these lines cross each other is the exact center of the guides, and we want to know if they are centered with the bore of the cylinder. If the guides are worn much, it will be in the center of them, where the greatest stress comes, and this cannot of course be changed except at great expense ; but it often happens that the cylinder shifts, and this can be remedied by a good machinist. We cannot give directions what should be done in such a case, but must leave the matter to be dealt with by every one to suit emergencies. The guides and cylinders of Corliss engines are supposed to be absolutely in line when new, and the method here illustrated is the one used to find out whether they are or not. Now suppose that we have our line centered exactly in the cylinder, the next thing we want to know is whether the shaft is exactly in the center of it. There are two ways to do this, and one of them is troublesome and expensive—the other is not. We show the easiest way. This is to drop the plumb line, *e*, at the exact center of the shaft, so that it just clears the cylinder line, using a try square on the end of the shaft with a blade long enough to reach the intersection of the two lines, so as to verify them ; try the square on both sides of the line, front and back, and the centre of the shaft will be accurately located. If the front end is low, the remedy is to raise it, of course, but before moving the shaft forward or back by the quarter brasses, the crank must be tried on the four quarters of its circle of revolution. This will show at once where the shaft stands with reference to the horizontal line, *a*, and the vertical line, *e*. Turn the crank over until it comes up to the cylinder lines, as in figs. 21 and 22. If the crank-pin is exactly midway of the collars with the line, it is right on that center. Now try it on the other center, and it will perhaps stand off ; if it does, the remedy is very plain. Now spring the cylinder line on one side so that the pin will pass it, and try the crank-pin on the vertical line, *e* (fig. 22); if it stands in toward the cylinder line, the center of the shaft is low

and must be raised. Try it on the bottom half
center also, and rectify it according to what the
line says. This method of lining up an engine
will cause the crank to revolve in a truly vertical plane, at exact right angles with the bore
of the cylinder. It makes no difference what

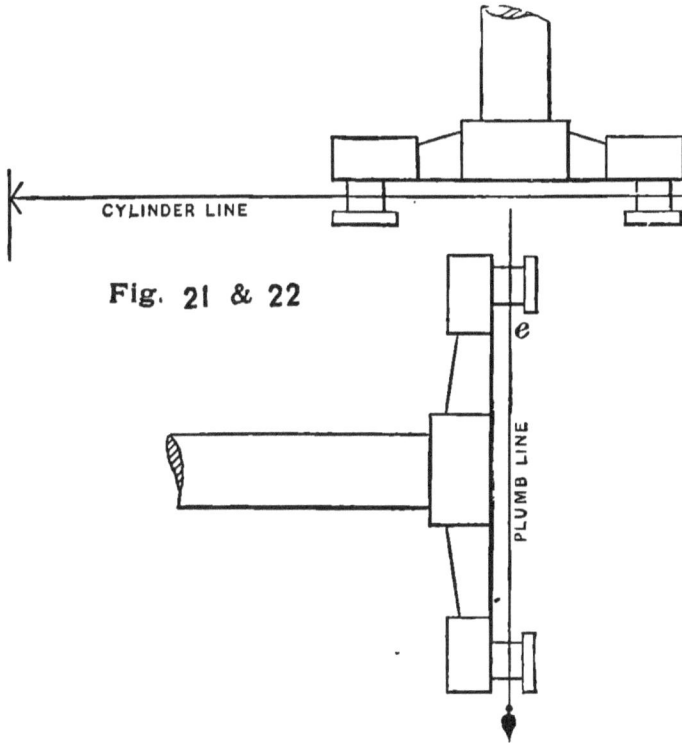

CYLINDER LINE

Fig. 21 & 22

e

PLUMB LINE

kind of an engine it is, the method is the same
for all, and any man of ordinary intelligence
can put his engine in exact line if he follows
these directions.

This is not to say, however, that all pounding will cease so soon as he has done so. We have fully adverted, in former chapters, to the causes of this, and need not repeat it. For the rest, time and experience can alone make an experienced engineer. No man can learn from a book exactly what to do with an engine to make it perform to the best advantage.

We have said nothing in this work as to the lubrication of an engine, but this is an important matter, and should be performed automatically. No one should use a squirt-can about an engine except for temporary use. There are various devices in market for feeding oil or grease to engines, and most of them are good. Sight feed lubricators are essential; no one now uses tallow or other animal fats. These last destroy cast-iron most rapidly. We leave this matter to the discretion of those in charge.

CHAPTER XIII.

MAKING JOINTS.

Joints about engines are, in the best practice, scraped or ground iron and iron, but in most machines they are made by interposing sheet-rubber, or patented compositions of it, which answer the purpose fully. There are a number of very good materials for this purpose on the market ; some engineers use one, some another ; while for metallic joints under high pressure the corrugated copper disks used are unsurpassed. These last can be used over and over again, and will not blow out or leak under any pressure. Where it is not possible to obtain these goods, a very good joint can be made with the wire-cloth used for mosquito-net frames. Cut this to the size required, and make a very thick paint with red lead and boiled oil ; daub this over the surface of the cloth and screw it up tight ; it will never leak or blow out, but it will be hard work to break the joint if it is suffered to remain for a length of time. If no cloth is handy, a single copper wire, say a scant eighth of an inch in diameter, will make a tight joint. Cut the wire the right length, stick it in a fire and heat it red-hot and plunge it into cold water. This will make it as soft as lead, so that it will flatten under the bolt pressure and fill all inequalities of surface. For face joints, like steam chest bonnets, hot or cold water pipes, heavy packing or drawing paper makes an excellent joint. Soak it in boiled oil and put it right on ; the heat will harden it into a parchment-like substance which is very serviceable. For permanent joints, like those in water-pipes under ground, or where they never have to be broken, a rust joint, so called, is the best. This can be made only where the castings are fitted for it in the design—that is, with a wide channel all around to receive the joint. A rust joint is made of fresh, clean cast-iron chips or borings which have no grease upon them. They are mixed with sal-ammoniac water and driven tightly into the space between the pipes, where the borings soon rust into a solid mass. Putty joints, so called, are used chiefly on cold-water pipes, about the feed pump, and may be used on hot-water pipes as well, if suffered to get hard before being put under heat and pressure. The putty is made of dry red lead and white lead mixed with oil, kneaded together to a stiff dough. It must be beaten with a mallet, and the stiffer it is the quicker it sets. It hardens into a mass as solid as a brick in time.

All these materials are only for exceptional use—that is, where the usual rubber or other gaskets can not be had, for these last are far more convenient than any just mentioned. In all cases where rubber joints are used they must be chalked or rubbed with a good black lead; this prevents them from sticking to the surfaces in contact. Joints should, in all cases, be made as long before their use as possible, so as to give them a chance to set before pressure is put upon them.

Packing the rods of steam engines is a simple matter, but simple as it is it requires judgment and good sense. Like every other duty about a steam engine, it needs to be properly done in order to work satisfactorily. Very many have an idea that the packing in a stuffing box must be jammed in as hard as it will go to prevent steam from leaking out, or what is just as bad, air leaking in when condensing engines are used. The reverse of this is true. There is no occasion to break studs and strip nuts on stuffing boxes to make a piston rod steam tight, but this very thing has been done by inexperienced persons. When a piston rod of any size can not be kept steam tight by moderate pressure on the packing, there is something wrong with the stuffing box itself, and this trouble in old engines, and in some new ones, too, will generally be found in the bottom of the stuffing box where the rod passes through the head. Too often this opening is made too large; in the case of old engines the piston rod has worn it oval by bearing on it. The remedy is to make a brass collar, or even of lead, which fits the piston rod nicely, and is one-eighth of one inch smaller than the stuffing box itself, or so that it is a loose fit. Put this in the bottom of the box, and a few turns of packing on top, moderately compressed, will keep the rods tight. As to the packing itself, use metallic packing where it is possible. There is no comparison between it and ordinary hemp packing used before there was any metallic packing. This last is always tight on good rods and runs with very moderate friction. It never needs screwing up or any other attention than to keep it in good working order. When metallic packing can not be had, an excellent substitute for it can be found in hemp gaskets braided firmly into a square, and thoroughly saturated with plumbago; that is blacklead. Do not make the mistake of using stove polish on gaskets because there happens to be plumbago in it. This quality is full of grit from the clay in it, and will badly score any rod to which it is applied. Some engineers use one thing and some another. There are various kinds of packing in market made from woven material, india-rubber, etc., etc., and engineers in large

towns can have a variety to select from. The
principal thing is, as has been said, to have the
stuffing box itself in good order; then very
little compression is needed. It would sur-
prise many who have never given the matter a
thought to see what resistance to motion a rod
two inches in diameter only, can offer when
packed tightly.

CHAPTER XIV.

Condensing Engines.

Thus far we have given attention wholly to engines which exhaust into the air, high pressure engines so called, or those which do not condense the exhaust. Condensing engines are sometimes called "low pressure" yet, but this is a term which is no longer applicable. It was used in the early days of the steam engine, when pressures were low, five and six pounds above the atmosphere. As a knowledge of boiler making increased, and higher pressures were available, the condensing apparatus was discarded as costly and cumbrous, and engines were made to exhaust into the air at higher pressures. To distinguish them from condensing engines, the terms high pressure and low pressure were used, but there is no longer any fitness in the appellation, for condensing engines will work at any pressure. The chief feature, then, of a condensing engine is that it exhausts into a vacuum instead of against the pressure of the atmosphere. Every one knows that this last plugs up the exhaust pipe with a pressure of 14.7 pounds upon every square inch of its area. All that the condensing engine does is to remove the plug and give a free exit to the exhaust. This is done by creating a

vacuum in a closed chamber called a condenser. Every one, even those with little or
no experience, knows this, but not all know
what a vacuum really is, or why and how such
a state of things is possible. We can not say a
vacuum exists, for it is not a thing. It is, in
fact, nothing; it has no existence. A vacuum
is simply absolute space, devoid of any fluid,
solid or gas. It can be obtained in two ways:
by mechanically pumping the air out of any
tight vessel, or by admitting steam to it and
throwing cold water in upon the steam. With
steam engines this is the usual way to obtain a
vacuum, and the philosophy of it is very easily
understood by what follows: Suppose we have
a cubic inch of water; that is a block of water
one inch square every way. Now, if we change
this into boiling water (212°), and let the steam
from it into a tight chamber one foot every
way, the steam will fill the chamber and be at
atmospheric pressure in it. Now, if we have a
pipe to this chamber, and run cold water in so
that it strikes the steam in a spray, the steam
will be condensed and fall to the bottom in the
form of water again, the air and condensed
steam falling together. Above this water
there is a vacuum more or less perfect; but to
make it an absolute vacuum we must remove
the water of condensation and whatever air

there is remaining. To do this a pump is necessary, and it is always present in condensing engines. It is called an air pump, and in action it removes the air and water of condensation from the condenser, leaving a more or less perfect vacuum, into which the engine exhausts. This operation goes on continually; the engine is always exhausting into the condenser, the cold water is always condensing the steam, and the air pump is constantly removing the water and mist held in suspension. To some this operation is a very complicated one, and many engineers say they can readily manage a high pressure engine, but do not know anything about a condensing engine. There is no reason why they should not understand the one as well as the other, for there is nothing in a condensing engine beyond the capacity of every intelligent man. The two evils to be guarded against in a condensing engine are air leaks and heat in the condenser. Look out for these, and there will be no trouble in maintaining a vacuum.

Since we have seen that a vacuum is absolute space, it is plain that if air leaks are present the vacuum will be impaired to just the amount that the air leaks in beyond the capacity of the air pump to remove it. If the condenser is not cold the steam will not be condensed quickly, and if the water of condensation and the vapor are not also removed, there will only be a partial vacuum, for the water of condensation is not absolutely cold, but at 120 to 140 degrees, and gives off vapor which also injures the vacuum. This is, in as few words as possible, the detail of a condensing engine, and it does not seem a formidable affair. There are two kinds of condensers in general use—the jet or absolute contact condenser, and the surface or indirect acting condenser. The first is simply a cast-iron vessel, usually round, as best adapted to resist the pressure of the atmosphere, for it must be remembered that the pressure on a condenser outside is many tons in the aggregate. (A condenser only 40" diameter and 72" long, under a perfect vacuum, has over 37½ tons total pressure on the outside, tending to crush it.) Into this vessel cold water is run through a perforated nozzle; when the water strikes the steam the latter is condensed and both the injected water and the condensed steam fall to the bottom of the vessel. The surface condenser is exactly the same as a common tubular boiler. The steam enters outside of the pipes (or flues) and the condensing water goes through them. The exhaust steam, therefore, does not strike the water directly, but is merely received upon a cold surface. and the water of condensation, only, falls to the bottom of the condenser; the condensing water passes away constantly through the pipes, or flues, and does not mingle with the condensing steam. This method gives absolutely pure water for the boiler feed, excepting only the foreign matters which may enter with the steam. Surface condensers are used chiefly upon ocean steamers, where they are indispensable, as they furnish fresh water for the boilers. In long voyages they are a necessity, and the greatest care is taken to avoid steam leaks, for this means a reduced supply to the boilers, Surface condensers also supply an immediate vacuum at the first exhaust of the engine. A circulating pump keeps cold water going through the tubes constantly, so that as soon as the exhaust steam strikes it it is condensed, and the main engines "take hold," as it is called. This is not always the case with a jet condenser, in which the vacuum is not very good for two or three revolutions. A vacuum is a vacuum, however obtained, and so long as one is produced that is the main thing. A loss of it is a loss of power, for the resistance of the atmosphere being removed from the exhaust side, the weight of it is added to the pressure on the piston. Thus, if the steam gauge shows seventy-five pounds, the actual or absolute pressure is 75x15, or 90 pounds. From this aspect we can readily see why engineers are sensitive about the condition of the vacuum, whether it is full or only partial.

The first mechanically made vacuum of which we have any authentic record is that of Torricelli, an Italian experimenter, and of Otto Guericke, a German experimenter. Which of these was the pioneer vacuum maker history saith not. Otto Guericke, of Magdeburg Germany, invented the common air pump, used in philosophical experiments, in 1654. He first tried, by filling a barrel full of water and pumping out the contents from the bottom, to obtain a vacuum above the water, but the barrel was not air tight and the experiment failed. He then made an ordinary metallic pump and obtained a vacuum. To show that air was a factor in the work of the world, and that we are surrounded by an atmosphere under pressure, he made a pair of brass hemispheres, which had a ground joint in the center and a cock in the stand at the bottom of them. He connected his air pump to these, and exhausted the air from the globe, and then hitched fifteen horses to an eyebolt in the upper hemisphere, but they were unable to pull the upper half off of the lower. This demonstrated conclusively that the atmosphere had pressure, for upon opening the cock and letting air into the hemispheres again, the balance was restored and the hemispheres fell apart. The first actual measure of the weight of the atmosphere is due to Torricelli, also about the middle of the seventeenth century.

CHAPTER XV.

This was by reason of a suggestion from the Duke of Tuscany, who, having dug a very deep well, proceeded to pump it out. He found, however, that he could not raise water over 32 feet, and he must have had a pretty good pump to do that. Not succeeding in getting water the Duke consulted Galileo, the famous philosopher who discovered the motion of the earth. This water problem was too much for him, and he gave it up. Shortly before Galileo died he gave the puzzle to Torricelli, who began to work with mercury as a basis of comparison of the relative weights of the pressure of the atmosphere. Now mercury is fourteen times the weight of water, and Torricelli argued that if the atmosphere would support a column of water 32 feet high (as it was proven it would in the case of the pump before referred to), it would also support a column of mercury one-fourteenth the height of the water, or 28 inches. To test this he took a glass tube, sealed at one end, and filled it full of mercury, displacing all the air therein. He then closed the open end with his finger and inverted the tube in a basin of mercury, when the mercury in the tube fell and settled, as he supposed it would, at 28 inches, leaving a vacuum in the upper part. Torricelli did not live long enough to do much with his discovery, but another philosopher, a Frenchman, Pascal by name, took it up and carried the experiments further. It occurred to him that if at the surface of the earth the atmosphere supported a column of water 32 feet high, at great elevations from the surface it would not support so much, because the atmosphere is rarer, or less dense; so he took the mercury column up on a high mountain.

Proof of Atmospheric Pressure.

At the top it registered only 25 inches, while at the bottom it was 28 inches. At other levels between the bottom of the mountain and its top he found varying registers on the mercury column, so it is established by inductive reasoning, supported by experiments, that at the surface of the earth water will rise in a perfect vacuum 32 feet, supported of course by the pressure of the atmosphere, for the vacuum itself has no power whatever; it is as stated previously, merely a space which offers no resistance, therefore, water or air rushes in to fill it.

No Power in a Vacuum.

It is important for engineers to know these facts because there are still a great many who are not aware of them, and suppose that a vacuum has some power in itself or that it has an existence as a force because it is measured on a gauge. This last is an error. Vacuum is not measured on a gauge, but atmospheric pressure is. We can not measure a nonentity and we must insist that engineers bear in mind that a vacuum is just that ; it is nothing but space. Space has neither weight, dimension, nor boundary; it is infinity.

Suppose the vacuum gauge shows 26 inches, what does that mean ? It does not mean that there is a space of 26 inches in the condenser which has no air in it, or that there are 26 inches of space in the cylinder which is a vacuum; it means that there is nearly an absence of air in the condenser, since the pressure of the atmosphere has forced the gauge index around to the 26 inch mark. Now, 26 inches represent 13 pounds air pressure, so we might just as well (perhaps better) mark vacuum gauges by pounds as by inches. The first vacuum gauges, however, were mercurial tubes on the Torricellian principle, and were marked in inches, so this system is still kept up. Suppose the vacuum gauge shows only 20 inches; then there is a partial vacuum only, for 20 inches are equal to 10 pounds only, and with the vacuum gauge at 20 inches there are four pounds pressure in the condenser, a dead loss to us, for we are working against so much back pressure when there should not be any.

Now, the best vacuum we can get with modern appliances and at the speed we run engines in these days is 26 to 27 inches, rarely the latter. This loss of one pound is due to the want of time to remove the last vestige of air and vapor, to the mechanical imperfections of our appliances, and to the fog or mist of condensation, which to a greater or less extent pervades all condensers, whether surface or jet. It is creditable that we are able to do so much as this, but the greatest enemy engineers have to contend with in maintaining a vacuum is air leaks, pure and simple. The joints about a condensing engine are almost innumerable, and each pinhole, even, contributes its quota of mischief. Leaks occur through bolt holes, through gaskets, through castings themselves. The chaplets used in foundries to support cores are very liable to be leaky. Look out for them, and daub them over thickly with red lead paint. Paint every part of the injection pipes thickly; keep all stuffing boxes of injection valves well packed and use every means you can think of to guard against loss of atmospheric pressure by leakage of air into the condenser. Go round to every joint you can reach with a lamp and hold the flame against it. If there are air leaks you

can sometimes hear them, but when too small
to be heard they can be seen, for the flame
will be forced in toward the leak. Keep the
foot valves in perfect order, and the air-pump
bucket as well, both must be as near air tight
as possible. Remember the adage: *"Nature
abhors a vacuum"* and will fill it in an incredibly
short time if she is not prevented.

If, when the engine is working well otherwise, the vacuum begins to fall, so to call it,
give more injection water. Sometimes steam
from the boiler is hotter than at others, the
water in the boiler falls and the steam is superheated; that calls for more injection. If the
vacuum is still poor try the condenser by hand
and if it is warm and getting warmer, and no
amount of injection water will keep cool, see
if the foot-valves seat properly. If they are
cocked ever so little the water of condensation
can not get out, and by lying in the bottom of
the condenser kills the vacuum with its vapor.
The injection pipes may also be stopped. In
fresh water, eels often get drawn into the pipes
and stop them, when the water is drawn from
ponds; in rivers and streams, weeds, and also
fish, get drawn across the strainer and prevent
the water from entering. Every injection pipe,
whether on sea or on shore, should have a
steam pipe let into it for use in emergencies,

with a nozzle pointing toward the source of
supply. It may save a long stop for cleaning
the water pipe.

A good vacuum is worth 13 pounds of steam
in the boiler, and the feed water is heated to
120 degrees without charge for the same. All
it costs is the extra machinery needed to obtain
it, an air pump, injection valve, and pipes and
details. This once paid for is a small expense
to maintain, so that for a term of years the
outlay for a condensing engine is soon made
up in the decreased cost per horse power per
hour. That condensing engines are not more
frequently employed is due to the belief on the
part of the steam users that they are complicated, costly to maintain and hard to manage.
The exact reverse of this statement is the correct one.

Pumps.

The popular idea is that a pump has something to do with raising water or oil, or mo-
lasses, or any other fluid it may happen to be
at work upon, but this is a gross error, first
pointed out in the pages of The Engineer.
By reason of this view, persons who run pumps
are very often troubled about the water which
comes to the pump, and, in case of failure of
the pump to act, they examine into the condition or connections to the water, as if these

had something to do with the difficulty. The
only thing which can prevent a pump from
working is air, and air leaks on the suction
side and force side, so-called. Actually there
is no suction side, neither is there any such
force exerted as suction. It is a term invented
and applied long before we knew what atmospheric pressure was, or recognized its great influence in the work of a steam engine. The
sole office and function of a lifting pump—so-
called—is to remove the air from the pipe
which conveys the water to the pump. When
this is done water flows to the pump by the
pressure of the atmosphere outside upon it,
forcing it up to the pump chambers. If the air
is completely exhausted the water enters freely
and the pump is said to work well. If it is
only partly exhausted the water flows slug-
gishly, and the pump works badly. If we need
proof that a pump has no direct effect on the
water itself, we can attach one to a pipe 36 feet
long. The water will then rise in the pipe for
32 feet, but after this occurs we may work the
pump for all time and not get a particle of water
through it. The reason of this is that the atmosphere will not support a column of water
over 32 feet in height, and therefore the pump
has no effect upon it.

CHAPTER XVI.

SUPPORTING A WATER COLUMN BY THE ATMOSPHERE.

Now there are many who do not understand clearly what is meant by the atmosphere supporting a column of water. They see a pipe

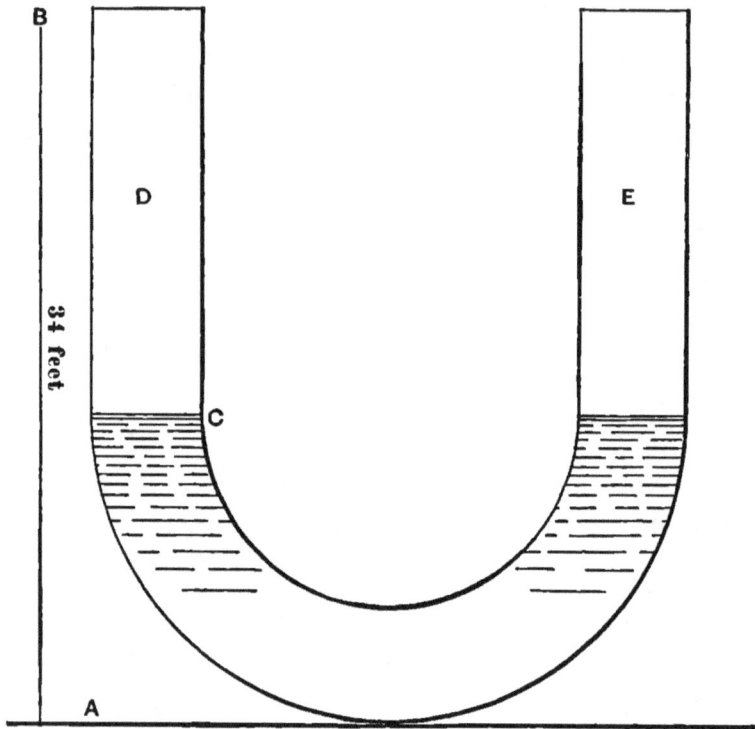

Fig. 23

full of water—a stand-pipe, for instance—which is merely a long tank upon end; or they see a tank at a railway station, and understand that these are not supported by the atmosphere, but are merely reservoirs which have been filled up. Where, then, is the difference? The diagram appended will make this plain. This is a U pipe, 34 feet long from its base at *A* to its top *B*. Now, suppose we fill this pipe up to the mark *C*, or any other mark equal to half the capacity of the pipe, and attach a pump at *B*, keeping it air-tight. When we exhaust the air from the arm *D* the pressure of the atmosphere in the arm *E* will drive the water up in *D* say 33 feet, if there is a perfect vacuum, and the water will stand there just so long as the vacuum is maintained, no longer, unless there is a valve at the bottom to prevent its return.

If the pump is stopped the water will fall to its level again, because air gets in through the pump and restores the balance. If there was no vacuum the water would stand at the same height in both arms, because there is just as much atmospheric pressure on one side of it as on the other side, and water being heavier than air, finds its level. This is all the mystery there is in the operation of a pump of any kind whatever, and aside from the mechanism which drives it if a well pump will not perform as it should, the reason can be generally found on the suction side so-called ; the plunger or bucket does not remove the air so that the water can get in. It is easy to see, then, that a pump demands the very best workmanship in all its interior working parts. The barrel should be smooth and true if a bucket works in it, and the packing of the same should be tight. If it is a plunger-pump, where the plunger is clear of the barrel, the stuffing box should be long, well packed, and the plunger itself should be true and work true in its path. Suspect every joint on the suction side of leaking, and if there are screw bolts which go through castings on the suction side suspect them also ; a good deal of air can get in through a very small opening. If there are many of these the net result detracts from the work of the pump. Friction of water is another element against a pump. This in the aggregate is very great in long pipes and in tortuous passages. Elbows and rough castings can take off much from the efficiency of a pump where the water has to be forced to it for long distances by the atmosphere, and this must be considered in the erection of any plant. There are only 14.7 pounds pressure on the water outside to get it where we want it, and this only when we have a perfect vacuum in the pipes ; with an imperfect one we have much less pressure. All this relates to what is known as the suction side of a pump, but, on the forcing side it is just as bad if the pipes run indirectly. All feed pipes should go as straight as they can be run to the boiler, but sometimes it is not possible to do this, and loops and vertical bends are made in them. Air collects in the tops of these bends and stops the water quite as effectually as a block of wood could. It lies on top of the water as wood floats on it, because it is very much lighter, and is compressed so much by the action of the plunger that it resists the main flow of the current, and the water surges back and forth in the chambers. This is fully shown in an air chamber, which is a well known adjunct to pumps, both single and double-acting. If metal valves are used for the lift or force sides, they should be carefully examined, from time to time, to see that they are tight on their seats, and lift squarely, and seat fairly. An unsuspected source of trouble is often found in the seats of valves. These last are brass bushes driven into cast-iron chambers. Sometimes these chambers are bored out for the valve seats, and very often they are not, but taken as they come from the foundry. In work of this character it is not uncommon to find leaks. The seats also work loose in the castings, and leak from that cause. Another difficulty with pumps is found in the lift or rise given the valves. Quick working pumps require very little lift to the valves on either side, but the most should be given on the force side. Divide the diameter of the valve by four; this will give a lift equal to the area of the opening in the valve seat, which is all that can be delivered to the pump barrel. A two inch valve, then, should lift only half an inch, and even this will be found too much in some cases. Plunger pumps that run at high speed, or over 100 feet per minute, are very apt to pound violently and make a great deal of noise. This can be overcome wholly by simply coning the end of the plunger to an angle of 30 or 40 degrees. Put the plunger in a lathe and bevel the end off, and there will be no more pounding. The reason for this is not easy to find. Pumps are still used in many places for feeding boilers, but in a majority of cases injectors are used. These last are simply managed, and the fullest directions are sent with them by the manufacturer. If they are followed implicitly there will be no trouble, but if persons undertake experiments on their own account they must not blame the apparatus.

These are succinctly the principles governing the action of condensing engines and the pumps by which they are worked. All pumps act upon the same principles as those previously alluded to. Whether the detail which exhausts the air from the water supply pipes is a scroll, a screw, or a fan attached to a shaft and rotated by it, as in a centrifugal pump, whether it is a simple bucket or a plunger, the fact is the same: the air must first be removed before any water can get to the pump, and the special detail, the fan aforesaid, or the bucket or plunger which forces the water out of the pump chamber has no direct influence upon drawing the water itself. It may be that we have reiterated this too often, but we think not, in view of the fact that we were told quite recently by a person in charge of a pump that the suction valves were so heavy that the plunger could not lift them. It is very hard to get rid of notions and ideas; the more erroneous they are the more difficult it is to abandon them. This is our apology, if any is needed, for insisting upon the facts laid down as regards the action of pumps. Also, let us say here, that in previous chapters we have stated that water would rise only 32 feet in a pipe in a perfect vacuum. We should have said in a working vacuum, which is far from being a perfect one. The mean pressure of the atmosphere within its known limits is 14.7 pounds per square inch, which corresponds to a column of mercury (supports it) 29.9 inches high, or will support a water column 33.9 feet high at the sea level. These are the exact figures, but we have all along in this work preferred to deal with every day results and figures, rather than submit mere cut and dried recitals of tabulated details.

Dismissing the steam engine and its belongings, with the bare review of its functions and management which has been possible in the assigned limits of this work, and assuming that we have a new plant to start for the first time, let us mention some details that are of great importance.

CHAPTER XVII.

STARTING A NEW PLANT.

New engines and boilers should be started with great care; this statement applies particularly to the boiler. If the latter is large, the fire under it should be started at least three days before the boiler is actually needed for work, and the fire should be very small indeed at first. For the first day no attempt should be made to raise steam, and the fire should not be urged in the least. The water should be allowed to get "hand-warm" only, and be kept at this temperature for twenty-four hours. The reasons for this must be apparent with very little thought. Everything is cold on the start, and all the dimensions will be greatly changed by heat, and unless great care is taken at the outset much injury can be done to the brick work setting and the boiler itself.

For the second day the temperature may be increased to nearly the boiling point, but the fire should not be driven. The furnace doors must be kept shut all the time, and the ash-pit doors also, the amount of draught and of fuel being governed so as to keep the boiler from making steam. On the third day the boiler may be allowed to make steam, but the pressure must be brought up gradually, and the fire upon no account forced. The furnace doors must be always kept closed as before. As the pressure rises above the atmosphere, open all the steam connections and allow the steam to warm the pipes thoroughly before putting greater pressure upon them. Do not close any valve with a rush when the pressure rises to the working point.

The boiler should be full of water on the start, three full gauges, so that while the pressure is still low the boiler can be blown down —through the blow-cock—to get rid of all the rubbish that has accumulated in inaccessible corners. Open the blow-cock steadily, not with a twitch of the handle, and blow down to two gauges. This should not be done until a few minutes before starting the engine; the feed will not be needed for a few minutes then, and in that time all the feed-pipe connections will warm up and expand equally.

Try all movable joints, handles, cocks, safety valves, everything in short, to see if they work properly, and examine every valve and stuffing box personally to see if they have been packed properly. Look carefully to all the joints under pressure, and do all this before a working pressure is raised; keep up this inspection from time to time as the pressure increases. On starting the engine, open all the cylinder cocks to blow out the condensed water which has accumulated in the pipes and cylinder. This is imperative, not only to get rid of the condensed water, but to blow out the sand, chips and minute filings, that can be removed in no other way. These have accumulated in the engine while it was being built and erected, and in no other way can they be so effectually removed. Move the live steam valves, so that steam is blown through both ends of the cylinder for the purpose mentioned.

Before turning the engine over the center for the first time make absolutely sure that everything is clear ; give the engine steam easily, and run the crank over on the three-quarter position; then give steam the other way, if there is hand gear which admits of it, and drive the crank back again. Do this carefully, and before the engine is finally allowed to pass the center shut the throttle entirely, so that if anything is wrong or anything carries away, the mischief will be confined to one stroke. The flywheel will carry the engine over the center.

An engine should be started the first time under very moderate pressure ; five pounds should be enough if the engine is properly made. No power is needed, and the only points to be established are, whether everything is in apparent good working order.

If possible, do not lace any main belts until after the engine has been tested. There is no knowing what may have to be done, for mistakes are possible to all until the engine has been tried. If indicator attachments are on, take friction diagrams at this time with the unloaded engine, and see what it requires to move itself. Do the same when the main belts and shafting are on. without the machines, and valuable data will be had for future reference. As to the engine connections, the main bearings, crank-pin, and cross-head end, should be left perfectly easy. If they thump slightly, it does not matter when the engine is running slowly. Thumping from loose connections is very different in sound from pounding for want of proper adjustment, and the careful and experienced engineer will detect the difference at once.

In all that has been said, we have endeavored to inculcate the idea that, above all other things, the most watchful care and supervision is needed on first starting a new engine and boiler. On such occasions a tremendous change is introduced. Cold metal is made hot, and, in this transition alone, inconceivable force is generated. It is none the less powerful because it is invisible, and makes itself known only by rupture. Boilers are made leaky by careless handling on the start which were perfectly tight and well made, and strains are set up within them by forcing them, which materially affects their life. The same is true of the brick-work, if the boiler is so set, and it is for this latter, primarily, that we advised three days moderate heating of the boiler upon starting it. It takes, or should take, a long time to heat a brick wall alike—so that it all goes together, and three days is none too long.

If these directions are followed, properly built engines and boilers will perform well from the start. There will be no running back and forth to the shop, or calking leaks, re-making joints, or any sort of fuss. There will be that harmonious straight-away condition of affairs which mark the difference between a man who knows his business and one who does not.

From what has been said in preceding pages it is apparent that to be a successful engineer requires care and skill of the highest quality. The attention necessary to keep a steam plant up to its best condition all the while must be unremitting, otherwise great loss results. It does not follow from this that an engineer should be hopping around from engine room to fire room, or running here and there with a squirt can, or in a fuss generally; what we mean to inculcate is that an engineer should keep the run of his plant in his head at all times, and not suppose things are all right because no accident has happened. Accidents never happen to careful men ; they only happen to persons who suppose instead of knowing, as far as human foresight can go. Mysterious boiler explosions, mysterious flywheel bursting, mysterious anythings about steam engines, could, if all the facts were known and the naked truth were told, be traced to a condition of things previously known to some one which was willfully neglected. "Let well enough alone," is a good maxim in an engine room, but this does not mean that bearings are never to be examined, boilers never cleaned, or never examined for defective braces, and the whole routine of an engineer's duties neglected. For ten hours daily, at the least, an engineer must keep watch of his engine and boiler, for things go wrong when they are least expected to. In a factory where hundreds of people are employed, a very small matter to an engineer may precipitate a panic which will cost many lives, and it is for him to see that it does not occur through his carelessness. We were in an engine room—fire room, rather—once when a rivet blew out above the water line, and made a great fuss. A youth who was in the place

started for the door, shouting that the boiler
had burst, but he did not get far enough to
frighten others before he was caught by the
collar and a little advice given him that was of
service. When a rivet blows out it is a simple
matter to whittle a pine plug and jam it in the
hole, either above or below the water line, and
it is not a bad idea to have plugs handy for this
purpose. It is not uncommon for rivets to
blow out.

Another point that an engineer should bear
in mind is that the engine is upon no account
to be stopped in working hours, unless it goes
to pieces, direct orders are given, or danger to
life and limb is imminent. No engineer should
stop a factory engine where goods are turned
out by the piece, or by the yard, or any other
quantity, for a hot bearing, or because some
detail of the engine will be ruined if kept running. The cost of most details of an engine is
slight, but if the detail costs a hundred dollars
it is better to lose it than a thousand dollars'
worth of work, or two hundred dollars' worth
of time. This is particularly the case in places
where power is sold to tenants. Every revolution
of the engine means some fractional part of a
dollar to them, and the stopping of an engine
for some trifling, or possibly serious, expense
to the landlord, might mean ruin to a tenant,

who would, perhaps, depend upon that very half hour to complete a contract in a given time. Upon trifles, as we call them, very great events depend sometimes.

We repeat again, never stop an engine in working hours except for the direst necessity. Also, never start an engine after it has been stopped without a direct written message, or direct personal notice, from the man in charge. Suppose nothing. It is a serious business to neglect either of the precautions above mentioned.

CHAPTER XVIII.

Water-Tube Boilers.

Now that water-tube boilers are supplanting fire-tube boilers, both for stationary and marine work, it is important that an engineer should know some of their chief features, and the reasons why they are driving out fire-tube boilers. These are, broadly, their immunity from disastrous explosions (there being no shell and but a limited quantity of water in them), their economy of maintenance, both in running and in upkeep, their accessibility for cleaning and their high efficiency as evaporators.

Boiler Explosions.

It is not asserted that no water-tube boiler has ever been ruptured as to its tubes, but it is asserted that no explosions like those of shell boilers—that is, fire-tube boilers—can be traced to water-tube boilers. The reason is, for one thing, that there is no shell of large diameter on water-tube boilers, and, for another, that the rupture of a tube acts like a safety valve—in a certain sense, and releases but a small quantity of water compared to the total water content of the boiler and compared to the water content of fire-tube boilers.

The idea that when a steam boiler is full of water it is in no danger of explosion is an absurd one and no longer entertained by intelligent engineers. The more water there is in a boiler which is in a condition to explode the greater the danger, for it is the large body of heated water giving up its stored energy of hundreds of tons which causes the terrible destruction when a fire-tube boiler explodes. The moment of explosion of a boiler's shell is small and its duration short when above the water line, but if the rupture occurs below the water line then the total energy of the heated water is directed to the injured part and destruction of the whole plant follows.

It is very similar to the ignition of a given quantity of gunpowder unconfined and a similar quantity enclosed in a tube. The so-called " mysterious " boiler explosion, which is often reported in the daily papers, is not mysterious to some one who was about it or in charge of it at the time and who knew of its condition, but refused to repair it.

Economy of Maintenance.

The water-tube boiler, as compared with the fire-tube boiler, is far more economical in its

freedom from costly repairs. The chief parts
which require renewal in a water-tube boiler
are the tubes nearest the fire, but with proper
management they will last about ten years.
When they do require to be renewed the expense is very small indeed per horse power of
the engine driven, and the time required, which
is also part of the cost, is scarcely worth mentioning. Water-tube boilers are in action to-
day which have not cost one cent for repairs of
any kind whatsoever after many years' use.
The New York Steam Company has 14,000
horse power Babcock & Wilcox Company
Water-Tube Boilers running night and day for
several years, the cost of repairs has been three-
quarters of one cent annually per horse power.
Other boilers of the same type have been in
constant use day and night without costing one
cent for repairs, so it is easy to see that, as
compared with fire-tube boilers, the water-tube
is the cheapest to run.

EVAPORATIVE EFFICIENCY.

Steam boilers in these days are rated by their
ability to turn water into steam, and the term
horse power cannot be properly applied to
them.

Regarding the " power " of steam boilers,

the word is misapplied, but it is still used by reason of custom, and because there is no other popular term to express a boiler of a given size. It is obvious that a steam boiler is merely a magazine of stored force which may be, and is, of varying power in accordance with the way in which it is used. A reservoir of water could not be said to be of 5,000 or any thousand horse power if its contents were directed on to a turbine wheel, unless it was also stated how long and with what volume and fall the water was used. Similarly, a steam boiler is of varying power for a given rating in grate and heating surface, according as its stored force is used. The rating of steam boilers is now expressed in terms of their ability to evaporate certain quantities of water into dry steam in a given time, and this is the only fair test that can be given. The purchaser then knows exactly what he is getting, and can use the steam in one hour or in ten hours. No questions enter into argument as to the amount of heating and grate surface; these things rest with the designer of the boiler, and it stands or falls by its performance. These last values, heating and grate surface, have greater or less significance, according to the disposition of them and their relation to each other. A square foot of heating surface in a boiler is of much greater efficiency in one place than in another. To merely state, then, that a boiler has ten square feet of heating surface to a horse power means nothing at all as regards its evaporative effect, and its performance cannot be accurately relied upon.

This will be clearer to non-technical readers when it is stated that a simple engine, having a single cylinder, should produce a horse power upon 30 pounds of water evaporated into steam at 70 pounds gauge pressure; a compound engine, having two cylinders and working at from 6 to 10 expansions, will produce a horse power for an expenditure of 20 pounds of water; and a triple cylinder engine, working at 16 to 30 expansions, should give one horse power for every 15 pounds of water evaporated into steam per hour, in all of the above citations. Now, the same boiler will supply all of these engines (in rotation) if the proper pressures for the work are carried, but the power developed is vastly greater with the high expansion engines than with the simple engine. Very much higher values could be given for high expansion engines, but the writer has taken the average. It seems plain, therefore, that the power of a boiler begins and ends with its ability to evaporate certain quantities of water in a given time.

Furthermore, the evaporative power of boilers depends largely upon the amount of coal burned upon the grate in a given time, so the power of a boiler of certain dimensions can be augmented over its normal power by using artificial draught of one kind or another, air driven in directly by a fan, or air drawn in by induction, as with a jet, or with the exhaust turned into the chimney.

Take the case of a locomotive; under the stimulus of the exhaust, a locomotive of say 1,200 square feet of heating surface and 18 square feet of grate surface will develop 600 horse power, but the normal capacity rating of a locomotive boiler under stationary boiler rules would be only 120 horse power. Each pound of coal boils off so much water into steam; with forced draught rather less per pound of coal than with natural draught, but since 75 pounds of coal are burned in the same time (per square foot of grate) that 15 pounds of coal are burned by natural draught, nearly four times the amount of water is boiled into steam in a given time.

The fact that high powers can be obtained from boilers of a given heating surface is well shown by fast yachts and by torpedo boats. A high speed steam yacht built last year has a boiler of only 1,200 square feet of heating surface, but this boiler has been worked up to over 600 horse power with quadruple engines and forced draught of great intensity. As regards this last, the punishment that a boiler will stand without giving up the ghost incontinently is astonishing, and water-tube boilers seem specially adapted to this method of driving them. Fire-tube boilers, especially those of the vertical type, are the least economical and efficient of their class, particularly when rated by the square feet of heating surface they contain. The tubes of plain vertical boilers utilize but from one-half to two-thirds of their total surface, for this portion is the only part reached by the water, with an exception in the case of submerged tubes. Suppose, for example, that a tube is four feet long above the fire box, the water would then be carried in it for only thirty inches of its length, the remainder being for steam room. A vertical tubular boiler having forty 2-inch tubes four feet long has a nominal heating surface in the tubes of 83 square feet, but owing to the defect of its type it has actually but 54 square feet that is of any value as steam generating surface. This is not the case with water-tube boilers of any other type, for the entire tube surface is directly in the fire, and exposed to an equal temperature all over. The average evaporation and consequently the efficiency per square foot of grate and per pound of coal is much higher in the water-tube boiler than in the fire-tube. Established records of one type of which there are larger batteries installed, greater aggregate horse power and longer in use than any other—the Babcock & Wilcox Company boiler—show that the evaporation runs from 10.94 pounds of water per pound of combustible to 11.84, and in one case reached the quantity of 12.42 pounds per pound of combustible. Even under "actual conditions," by which is meant every day work, with coal as it came and the boiler as it was, clean or dirty, the evaporation averages ten pounds.

Contrast this with the average evaporation of the average fire-tube boiler of all types, and it is easy to see why the water-tube boiler gains in favor.

In comparing these two classes of boilers— fire-tube and water-tube—general accessibility is a feature of importance, and in this respect the water-tube boiler surpasses the fire-tube.

Horizontal fire-tube boilers are very difficult to keep clean, some parts being impossible to reach, beneath the lower course of tubes, for

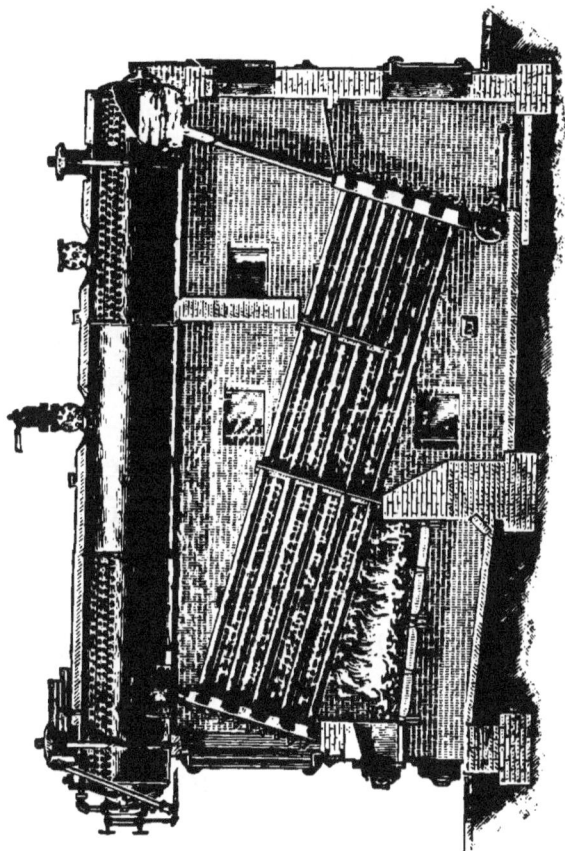

FIG. 24.—Babcock & Wilcox Water-Tube Boiler.

example, but in inclined water-tube boilers the tubes can be opened from end to end.

MANY KINDS OF WATER-TUBE BOILERS.

Broadly, all boilers which have the fire outside of the tube and the water inside of it are water-tube boilers, but there are very many kinds of them; that is, the arrangement and disposition of the heating surfaces varies greatly; also the proportions of grate to heating surface and heating surface for a given evaporation, but it is proper to say that there are more inclined tube water-tube boilers in use than any other. They are fast superseding fire-tube boilers in electric-lighting stations, electric railways, water works, and sugar houses. The oldest and best known of this type is the Babcock & Wilcox boiler, and it is shown in Fig. 24.

The horizontal tube water-tube boiler is here represented by the Roberts Safety Water-Tube Boiler, Fig. 25, and it is also the oldest and best known of its type.

In this type of boiler the water is delivered by the pump into two feed heating coils, one on each side of the drum, which abstract a good deal of heat from the gases that would otherwise pass up the smoke-stack. From these coils the water passes into the drum. This heated water is then taken into the circulation and carried down through the " downflows " —the two large pipes on each side shown in the front of the cut—the same number being at the back end. From the " downflows " the water passes into the side pipes, one on each side of the grate bars, and up through the up-flow coils and into the drum. The upflow coils are directly over and form the crown of the furnace. The steam rises to the top of the drum and the water not generated into steam,

FIG. 25.—Roberts Safety Water-Tube Boiler.

but carried up by the steam, is sent down the "downflows" again by the circulation. The steam then passes through a spray pipe in the drum and out into two superheating coils—one of which can be seen above the fire-brick on the side of the cut. The superheated steam supply is taken from the terminals of these coils.

FIG. 26.—Watson Radial Water-Tube Boiler.

The sub-vertical tube type of water-tube boiler is well represented by the Watson Radial Water-Tube Boiler, Figs. 26 and 27.

This boiler is constructed wholly of steel plate and steel tubes, and is, therefore, free from strains caused by metals with variable ratios of expansion. In its present form any tube which may wear out by corrosion or use can be withdrawn readily in a short time, there being but one bolted joint, and that a small one, to break.

Construction: The tube system is in the form of a cone, inclined over a central furnace, the gases and heat from which are diverted between the tubes by a baffle-plate in the upper part directly under the smoke-tube. The tubes are fastened in the tube-sheets by expanding them in the lower sheet and expanding them in the upper tube-sheet. The tube-sheets are inclined, as will be seen in the sectional engraving, so that the tubes are at right angles with them.

It will be seen by inspecting the sectional engraving that the tubes are very widely spaced at the bottom, and converge at the top; this greatly facilitates the circulation of the gases and also breaks them up so that combustion of them is assured. The ash-pit doors are close

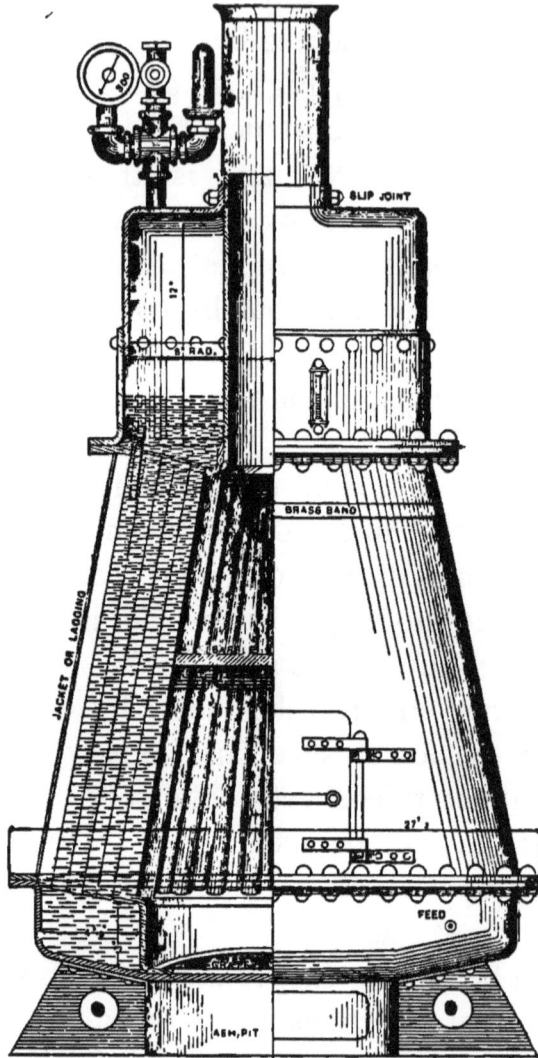

FIG. 27.--5 H. P. Watson Launch Boiler.

Height over dome, 3 ft. 8 in.; weight, 400 lbs.
With exhaust in Stack will develop 7 H. P.

to the ground and can be opened or closed by
the foot; fires are easily cleaned and cinders
raked out, as all engineers can see.

Circulation: A feature of the Watson Radial
Water-Tube Boiler is its perfect, natural circulation from the moment a fire is started.
This is attained by the following means: A
solid air-tight sheet-steel diaphragm extends
from bottom to top between the outside row
of steam tubes and the circulating tubes, broken
as to continuity by the fire door only. Outside
of this diaphragm are the circulating tubes.
The action is as follows: So soon as heat
strikes the inner row of tubes the water in
them is driven up, and in obedience to a natural
law, flows outwardly toward cooler water. A
pot on the fire always boils in the centre first,
and in like manner the water in this boiler
follows the same law. As it rises in the inner
row of tubes it falls in the outer (or circulating
tubes) and the cold water is constantly dis-
placed by heated water. This causes the water
to circulate rapidly and the boiler gets hot all
over simultaneously. The advantage of this
is too obvious to need further comment, and
accounts for the rapid steaming of the boiler.

The bent-tube type of water-tube boiler is
shown in the engraving, Fig. 28, which

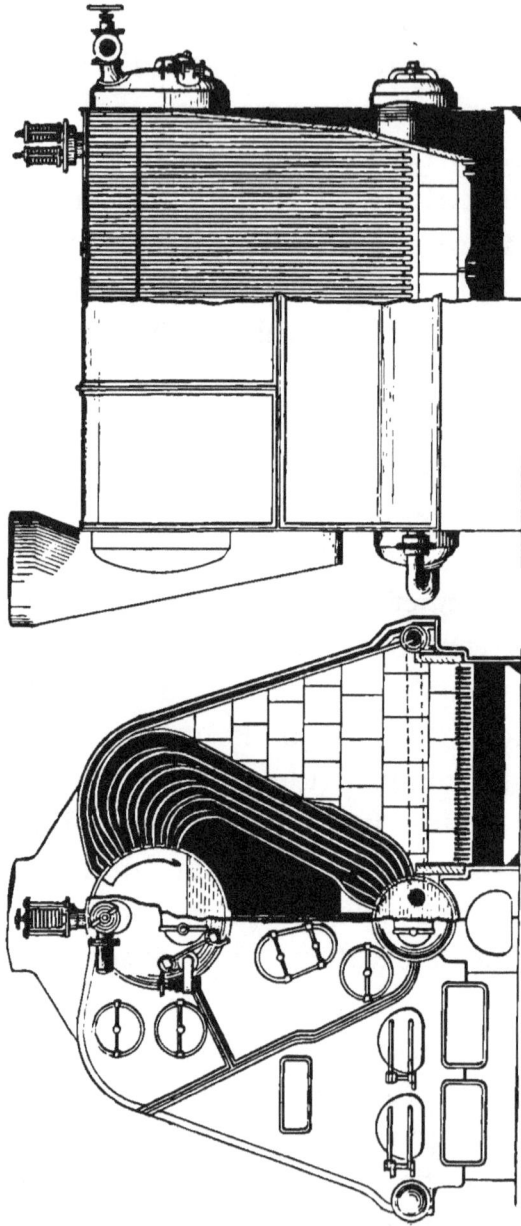

FIG. 23.—Thornycroft Water-Tube Boiler, "Daring" Type.

gives two views of the " Daring " type of
the Thornycroft boiler. The Thornycroft
" Launch " boiler, Fig. 29, is good for light

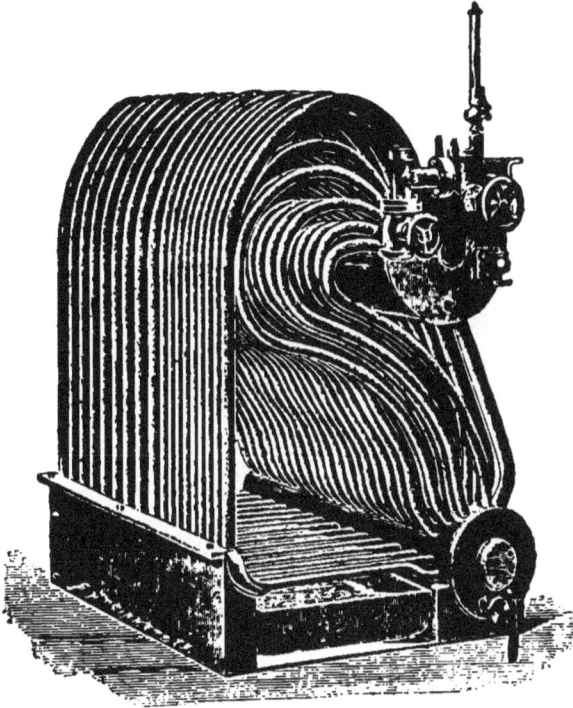

FIG. 29.—Thornycroft Water-Tube Boiler, "Launch" Type.

river work. It is provided with water fire bars,
which are kept so much cooler than ordinary
bars by the strong current of water rushing
through them that clinkers do not adhere, and
a great deal of the trouble usual in firing small
boilers is eliminated. Their weight is, of
course, exceedingly low.

Fig. 30 shows a side view of Babcock &
Wilcox wrought steel construction marine
boiler for 200 pounds pressure. All the pressure parts of the boiler are of wrought steel.

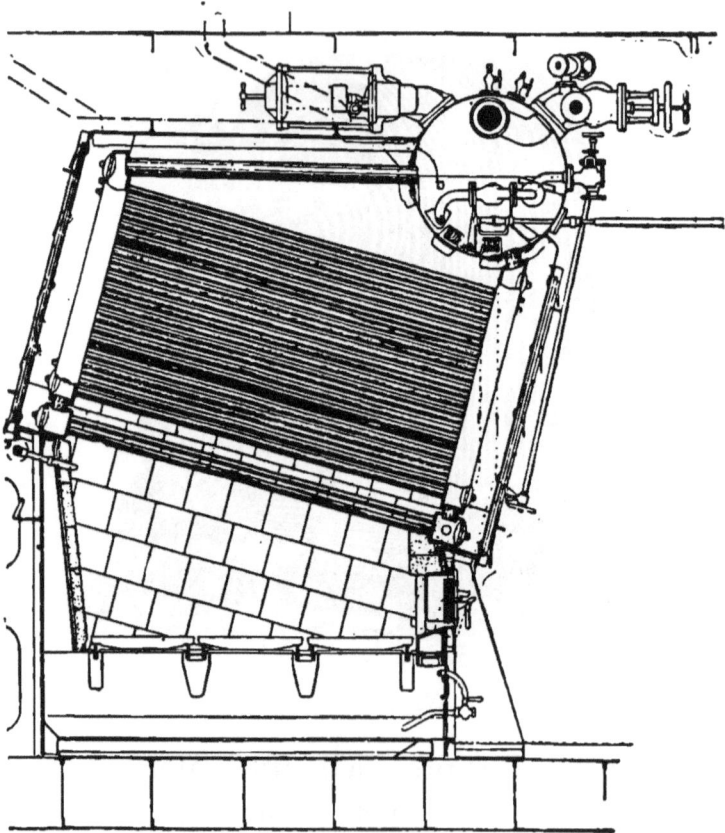

FIG. 30.—Babcock & Wilcox Marine Boiler.

Before passing from marine boilers, mention must be made of the well-known Bellville
type of water-tube boilers. This boiler has

been installed on some of the fastest cruisers afloat. See Fig. 31.

All of these boilers have peculiar features of their own and have different proportions, but it is not possible to give details of their construction, neither would it be proper in a work of this kind to express a preference for one type over the other. They are employed both in stationary and marine work, and, as may be seen from their details, are rapid steamers. With some of these boilers for marine work it is possible to generate 150 pounds of steam by natural draught from water at 35 degrees F. in twenty to thirty minutes, and all of them are constructed to stand 250 pounds of steam and upward. In order to pass Government inspection they must be tested to double that pressure, and it is easily seen from this fact that such boilers have great strength; all the materials are of steel and all the holes are drilled, and no plate under 60,000 pounds tensile strength is allowed to be used in their construction.

Why Water-Tube Boilers Steam Rapidly.

By looking at the engravings it will be seen that the water in these boilers is contained in tubes of small diameter directly in the fire, or in direct communication with exceedingly high temperatures.

A tube of 1-in. outside diameter and 30 in. long contains half a pint of water; the superficial area exposed to heat of such a tube is 94¼ square inches, or about the area of an 8 in. by 12 in. pane of glass. Now, half a pint of water on this surface is a thin film, and when exposed to the heat of a very hot fire, say 2,000 degrees F., is evaporated instantly, as one may say. By the plan of the boiler this water is in continual circulation, sweeping over all the highly heated surfaces, so that, virtually, the water constantly goes in at one side, or one end, as the case may be, continuously issuing at the other end as dry steam. All boilers are not fitted with 1-in. tubes; some types have much larger tubes, but their superficial area is correspondingly greater as well. Compare this action with the fire-tube boiler, where the water is anywhere from 3-in. to 6-in. deep on the tubes, and we have the explanation why the water-tube boilers are the most efficient, by which term is meant that the heat generated by the fuel burned is applied directly to the heating surfaces, which are, again, directly in the zone of the fire and close to it, so that the products of combustion have but short distances

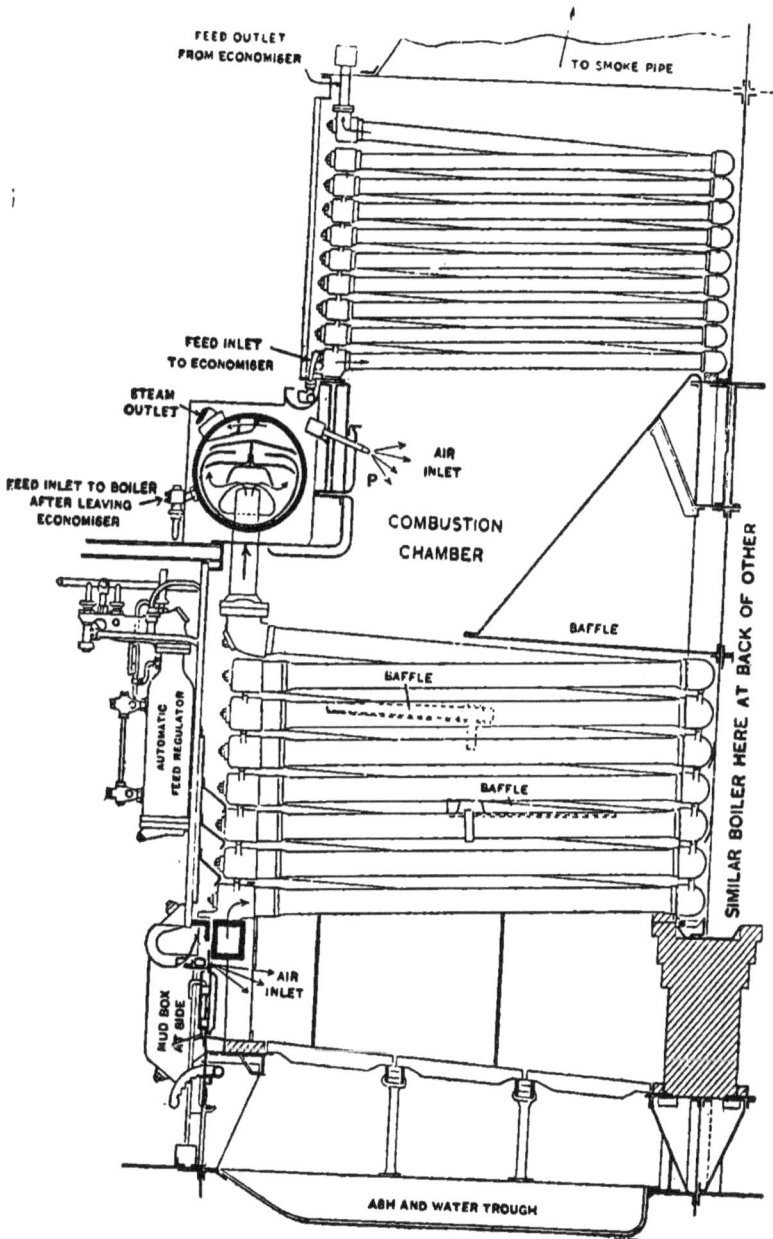

FEED OUTLET FROM ECONOMISER

TO SMOKE PIPE

FEED INLET TO ECONOMISER

STEAM OUTLET

FEED INLET TO BOILER AFTER LEAVING ECONOMISER

AIR INLET

P

COMBUSTION CHAMBER

BAFFLE

AUTOMATIC FEED REGULATOR

BAFFLE

BAFFLE

SIMILAR BOILER HERE AT BACK OF OTHER

MUD BOX AT SIDE

AIR INLET

ASH AND WATER TROUGH

FIG. 31.--Bellville Marine Boiler.

to travel to reach their work. This cannot be said of fire-tube boilers set in masonry.

TORPEDO-BOAT BOILERS.

From the facts just cited as regards water-tube boilers it is possible to get very high evaporation (boiler power, so called) from a very small and very light apparatus, and it is this quality which makes them particularly suited for torpedo boats and vessels akin to them (high speed yachts). As a rule it may be said that these boilers are less than half the weight of fire-tube shell boilers and far more compact. They have a very low center of gravity and for very considerable powers can be put under deck in a light draught vessel. The heating and grate surfaces allotted to the water-tube boiler are much less than the fire-tube boiler per horse power evaporation. This last quantity varies from 30 pounds of water for a common slide valve engine to 20 pounds of water for a two-cylinder compound engine, and to 15 pounds for a three-cylinder or triple expansion engine. It has been found in practice that five square feet of live heating surface per one horse power evaporation is ample for boilers of this type, and many of them show a horse power evaporation upon three square feet of live heating surface; the grate surface runs from one square foot of grate to twenty-five square feet of heating surface, to thirty-five, forty, and as high as fifty square feet of heating surface to one square foot of grate surface. Compare this with the allotment in fire-tube boilers for the same purpose, ten and twelve square feet of heating surface per horse power evaporation, and additional evidence is given of their relative efficiencies.

Mention has been made in previous lines of the high powers exerted by water-tube boilers of the torpedo boat type, and these are certainly phenomenal. It seems quite impossible to persons familiar with heavy, slow combustion, shell boilers that boilers of five and six thousand power can be put into a small vessel of say 130 feet length by ten feet beam, or width; far less superficial area than is contained in an ordinary city lot, for this last has 2,500 square feet area while the torpedo boat has little over half of it.

It will aid to a full comprehension of the apparent paradox when we consider the type of engine used and the steam pressures carried. The engines are in all cases high expansion engines—triple or quadruple stage—and run at high piston speed, 1.000 and 1,200 feet per minute being not uncommon. Now the boilers of these torpedo boats are only directly concerned with the high pressure or first cylinder of the system, because it exhausts into all the others in turn, and if it can supply the first cylinder, at say 250 pounds gauge pressure, the boiler gets credit, so to speak, for the power exerted by the other three cylinders. If, therefore, the boiler can manage the high pressure cylinder developing one thousand horse power per hour, it is furnishing steam (under the conditions) practically for three thousand horse power.

Again, take the case of a locomotive engine in common use. An average modern express engine has about 1,800 square feet heating surface; at the rating usually followed in stationary practice this would only give 180 horse power for the boiler under natural draught, but with the stimulus of the exhaust in the chimney a locomotive boiler will develop a horse power for less than two square feet of heating surface and supply two cylinders 20-in. by 24-in. at 300 revolutions per minute, when under natural draught it would not supply one at 150 revolutions per minute. Examine the grate surface of the express locomotive boiler and it will be found to have 30 square feet, a ratio of heating surface to grate surface of 60 to 1, or about half of what a stationary boiler would require for the same work. It follows, therefore, that high powered boilers of small size obtain their efficiency from their ability to generate large quantities of steam rapidly at very high pressures when under forced draught, the conditions being entirely dissimilar from those of stationary boilers.

MANAGEMENT OF WATER-TUBE BOILERS.

Under ordinary circumstances—that is with natural draught and slow combustion, the management of water-tube boilers is the same as that of any boiler, but since the tubes are small (and these control the amount of water exposed to the fire) it is necessary that they should be kept absolutely clean. This is particularly the case with the lower course of tubes, or those nearest the fire. These take the most intense heat and scale must be carefully guarded against. If the water is at all bad, or hard* it is essential to give strict attention to this detail, for unless these tubes are kept free trouble of a serious nature is certain to occur. It is very easy to say that every one knows this, and it is therefore superfluous to mention it, but it does not always follow that what every one knows every one attends to.

*See Collett on Water Softening. (Spon & Chamberlain.)

The old saying that " want of care does more harm than want of knowledge " is particularly · applicable to all boilers, and to sum the whole case in a very few words, the careful man about a steam boiler is the one who never has any trouble. Every man in charge of a steam boiler is supposed to have sense enough to see that he always has water at the proper level before putting fire under it, but there is a great deal of laxity in this matter of firing.*

Not so much is said in the public prints about the smoke nuisance as there was a few years ago, and there is no question but that the agitation of this matter of smoky chimneys in large cities was a very good thing for steam users, for if it did nothing else, it caused firemen to attend more carefully to their work. Smoke prevention starts directly from the furnace door, that is to say, the fireman practically prevents it by not making it, but he has to do a good deal more work than in the days when he threw in all the furnace would hold and then took a rest for half an hour with a smoke pipe of his own in his mouth. Practice proves that the smokiest coal can be deprived of its terrors by judicious firing, " the little and often " theory; but it is harder on the man

* See Dahlstrom: The Fireman's Guide. (Spon & Chamberlain.)

behind the shovel. The writer has fired bituminous coal—a scoopful at a time—without producing smoke of any moment at all, but it is not a pastime to do it, and it must not be wondered at that mere flesh and blood rebels against " crooking the pregnant hinges of the knee " hour after hour at such work. It is also the most economical method of firing, and those who have been used to heavy firing are advised to try light as an improvement.

Firing a steam boiler, however, has to be done differently with every plant, or more properly, all plants cannot be fired in the same way, nor all grates in the same manner, but each one has its peculiarities which those in charge must discover. Cleanliness in boilers— inside and out—is as important as anything else about them; soot is a non-conductor, and it is not to be expected that a boiler will steam freely if the heating surfaces are covered with it.

If boilers do not steam freely when they have never given any trouble previously the cause may be found in a combination of several things. Sometimes the atmosphere itself is so " heavy " as it is called that the fuel will not burn or the gases combine with it, but it is more frequently the fault of dirty fires when steam is short than a heavy atmosphere. The ash pit, should be always bright, and one can tell at a glance whether the fire is doing its duty or not.

" Perfect combustion," so often claimed for this or that appliance, or smoke preventer, is impossible under the conditions prevailing in commercial steam making, because different volumes of air are required at every stage of the process. It is important to distinguish between weight and volume in combustion. A given weight may enter through a one-inch hole if it has velocity enough, but volume is required in order that the air and gases may be thoroughly mixed, but while we cannot hope for perfect combustion we can obtain fairly good combustion by careful attention to the fires.

CHAPTER XIX.

The Highest Qualities Demanded.

Finally, let me say, in conclusion of this work, that the duties of an engineer worthy of the name call for the highest qualities, and are not to be lightly undertaken, or held in low esteem. A man who stops and starts an engine is not an engineer, and has no pride in his business, because he knows nothing of it ; he does not wish to know any more than that opening the throttle lets steam into the chest.

But we should not be discouraged or careless ourselves because such men get places, to the exclusion of their betters. There are usurpers everywhere. Quack doctors abound, so do quack ministers and shyster lawyers. It would be quite as logical and sensible for skilled professional men of these classes to give up trying to rise as it would be for an engineer to follow the same course. Knowledge of our business is paid for always, but an engineer must know where to find the best market for his services, exactly as every other man must who has something to sell. A dentist, let us say, settles in a certain locality and does not thrive. He does not immediately accuse his profession as the cause of his trouble, but he says that there is no business in that place, and searches until he finds one that is better. "Knowledge is power" is as true to-day as ever, but there are some places that are better than others to sell it in.

If an engineer spares no effort to improve himself, and studies first principles so as to know where to look for the cause and cure of troubles never encountered before, he is a better man for a steam user than a mere stopper and starter who does not wish to learn. Somewhere there is a steam user looking for him, and it is the engineer's business to find a place where he is paid for his work. We need only look around us to see engineers who have good homes, are socially esteemed, and are bringing up families to be a credit to themselves and the State. These men started from small beginnings, and were careful, prudent and anxious to learn. They did learn, and that is why they thrived.

The Man Himself is the Factor.

It is not the business which a man follows that keeps him down or lifts him up ; it is the man himself in every case, and it is well to bear in mind that a man can not be an engineer, or a lawyer, or a doctor, or anything else, at a bound. Long service, patient waiting, disappointments, reverses, learning through them and learning by success also—all these have their perfect work. No faithful service is ever lost. If a steam plant is in perfect order and running lower than others in the vicinity be assured that if the employer does not see it, others do, and perhaps when we least expect it we may get a call to go elsewhere with manifest benefit. There are many things conducive to success in engineering, as in all other callings which mankind follow, and none of these has more effect in business intercourse than a pleasant address. Engineers are commonly supposed to be "rough men," but after living and associating with them for forty odd years, all over the United States, and of all classes, locomotive, stationary and marine, we have found fewer engineers of violent manners and rude bearing than we have in other professions. Some may feel that civil speech has little to do with success. It has everything to do with it, for as a rule, even if men are skillful in their special line, we will not encounter them if we can avoid it when they greet us roughly and are surly in their dealings with us.

LASTLY.

In these United States no man is above his calling or beyond it. If he is a man in all that the word implies, he is independent of circumstances and of conditions, and is always in demand. The trickster perishes by his own

sword. It does not take long to discover
whether men are honest or the reverse, and
once the verdict is given either way no one
can escape the consequences. Not merely
honest in the sense that he will not take what
does not belong to him, but an honest man in
a moral sense. And with this little sermon we
say farewell.

INDEX.

BOOKS FOR ENGINEERS.

STEAM-ENGINES AND BOILERS. An elementary text-book for young students. By Prof. J. H. Kinealy. Illustrated, 8vo, cloth.

THE WORKING AND MANAGEMENT OF STEAM BOILERS AND ENGINES, SHAFTING GEAR AND MACHINERY. F. V. Colyer. 12mo, cloth.

THE CORLISS ENGINE AND ITS MANAGEMENT. By Henthorn and Thurber. Illustrated.

FIREMAN'S GUIDE ON THE CARE OF BOILERS. Dahlstrom. 12mo, cloth.

THE SLIDE-VALVE SIMPLY EXPLAINED. By Tennant and Kinealy. Illustrated, 12mo, cloth.

LUBRICANTS, OILS AND GREASES. By I. I. Redwood. Illustrated, 8vo, cloth.

RICHARDS' STEAM-ENGINE INDICATOR. By Porter. Illustrated, 8vo, cloth.

THEORETICAL AND PRACTICAL AMMONIA REFRIGERATION. By I. I. Redwood. Illustrated, 12mo, cloth.

THE REPAIR AND MAINTENANCE OF MACHINERY. By T. W. Barber. Illustrated, 8vo, cloth.

QUICK AND EASY METHODS OF CALCULATING WITH THE SLIDE RULE. By R. G. Blaine. Illustrated, 12mo, cloth.

ALGEBRA SELF-TAUGHT FOR THE USE OF YOUNG ENGINEERS. By W. P. Higgs. 8vo, cloth.

DIRECT-ACTING PUMPING ENGINES. By P. R. Bjorling. Illustrated, 12mo, cloth.

SEXTON'S BOILER MAKER'S POCKETBOOK. Illustrated, 32mo, leather.

BOILER MAKER'S AND SHIPBUILDER'S COMPANION. By J. Foden.

MANAGEMENT AND WORKING OF STEAM BOILERS, LAND AND MARINE. By J. Peattie.

SPON'S ENGINEER'S TABLES.

SPON'S MECHANIC'S OWN BOOK. The book for every one.

WORKSHOP RECEIPTS (in five series).

Also books on Mechanics, Electricity and General Engineering. Catalogues free on application.

THE PRACTICAL APPLICATION

OF THE

SLIDE VALVE and

LINK MOTION

TO

Stationary, Portable, Locomotive and Marine Engines,

WITH NEW AND SIMPLE METHODS FOR PROPORTIONING THE PARTS,

By WILLIAM S. AUCHINCLOSS, C.E., Mem. A.S.C.E.

CONTENTS

USEFUL BOOKS

Barometer.—The barometrical determination of heights. A practical method of barometrical levelling and hypsometry, for surveyors and mountain climbers. By Dr. F. J. B. Cordeiro, U. S. N. 12mo, leather,   1.00

Dynamo.—Notes on the design of small dynamo, with complete set of drawings to scale. By G. Halliday. 79 pages, illustrated, 8vo, cloth, ... 1.00‡

Electric Bells.—A treatise on the construction of electric bells, indicators and similar apparatus. By F. C. Allsop. 131 pages, 177 illustrations, 12mo, cloth, ... 1.25

Electric Bells.—Practical electric bell fitting. A treatise on the fitting up and maintenance of electric bells and all their necessary apparatus. By F. C. Allsop. 170 pages, 186 illustrations, 12mo, cloth, . . . 1.25

Electrical Notes.—Practical electrical notes and definitions, for the use of engineering students and practical men. By W. Perren Maycock, E.E. 286 pages, illustrated, 32mo, cloth, ...75

Electricity.—Comparisons between the different systems of distributing electricity. By Prof. Henry Robinson. 8vo, paper, . . .80

Galvanometer.—A series of lectures on the galvanometer and its uses, delivered by Prof. E. L. Nicols, and used by him in his class at Cornell University. 112 pages, 76 illustrations, 8vo, paper, ... 1.00

CORRESPONDENCE INSTRUCTION . .

IN

ENGINEERING.

Steam Engineering,

Electrical Engineering,

Civil Engineering,

Mechanical Engineering,

MECHANICAL AND ARCHITECTURAL DRAWING,

Plumbing, Heating, Ventilation,

Chemistry, Metal Work, Mining.

THE

International Correspondence Schools,

SCRANTON, PA.

RUHMKORFF

INDUCTION COILS

Their Construction, Operation and Applications, with
Chapters on

BATTERIES, TESLA COILS AND ROENTGEN RADIOGRAPHY . . .

By H. S. NORRIE

❧ ❧ ❧

CONTENTS OF CHAPTERS:

183 pages, 12mo, cloth, 50 cents.

Manual of Instruction in
Ђard Soldering
WITH AN APPENDIX ON THE
Repair of Bicycle Frames
Notes on Alloys and a Chapter on Soft Soldering

BY HARVEY ROWELL

The flame, lamp, charcoal, mats, blow-pipes, wash-bottle, binding wire, chemicals, borax, spelter, silver solder, gold solder, oxidation of metals, fluxes, anti-oxidisers, oxidation of cases, the cone, oxidising flame, reducing flame, heat transmission, conduction, capacity of metals, radiation, application, the work table, the joint, applying solder, applying heat, the use of the blow-pipe, joints, making a ferrule, to repair a spoon, to repair a watch case, hard soldering with a forge or hearth, hard soldering with tongs, preserving thin edges, silversmith's pickle, restoring color to gold, chromic acid, to mend steel springs, sweating metals together, retaining work in position, making joints, applying heat, preventing the loss of heat, effect of sulphur lead and zinc, to preserve precious stones, annealing and hardening, burnt iron, to hard solder after soft solder. Tables of—specific gravity, tenacity, fusibility, alloys.

66 pages, illustrated, cloth, 75 cents.

For Soldering Receipts, Cements and Lutes, Pastes, Glues and such like, *see* WORKSHOP RECEIPTS.

SMALL ACCUMULATORS

How Made and Used

A Practical Handbook for Students and Young Electricians

EDITED BY PERCIVAL MARSHALL, A.I.M.E.

Contents of Chapters

I.—The Theory of the Accumulator.
II.—How to make a 4-Volt Pocket Accumulator.
III.—How to make a 32-Ampere-Hour Accumulator.
IV.—Types of Small Accumulators.
V.—How to Charge and Use Accumulators.
VI.—Applications of Small Accumulators, Electrical Novelties, etc. Useful Receipts. Glossary of Technical Terms.

80 pages, 40 illustrations, 12mo, cloth, 50c.

THE MAGNETO-TELEPHONE

ITS CONSTRUCTION,

Fitting Up and Adaptability to Every=Day Use

BY NORMAN HUGHES

CONTENTS OF CHAPTERS

Some electrical considerations : I.—Introductory. II.—Construction. III.—Lines, Indoor Lines. IV.—Signalling Apparatus. V.—Batteries. Open Circuit Batteries. Closed Circuit Batteries. VI.—Practical Operations. Circuit with Magneto Bells and Lightning Arresters. How to Test the Line. Push-Button Magneto Circuit. Two Stations with Battery Bells. VII.—Battery Telephone. Battery Telephone Circuit. Three Instruments on one Line. VIII.—General remarks. Index.

80 pages, 23 illustrations, 12mo, cloth, $1.00. In paper, 50c.

THE
FIREMAN'S GUIDE

A Handbook on the Care of Boilers

BY KARL P. DAHLSTROM, M.E.

CONTENTS OF CHAPTERS

8vo, cloth, 50 cents.

THE CORLISS ENGINE.

By John T. Henthorn.

— AND —

MANAGEMENT OF THE CORLISS ENGINE.

By Charles D. Thurber.

Uniform in One Volume. Cloth Cover; Price, $1.00.

Table of Contents.

Third Edition, with an Appendix.

7

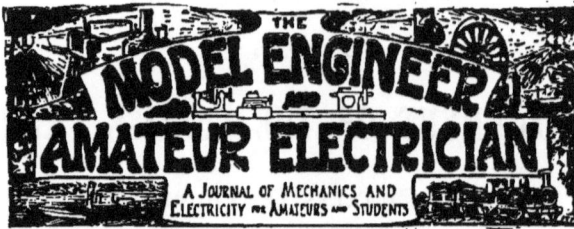

PRACTICAL HANDBOOK ON

Gas Engines

With Instructions for Care and Working of the Same.

BY G. LIECKFELD, C.E.

Translated with permission of the Author by

GEORGE RICHMOND, M.E.

WITH A CHAPTER ON OIL ENGINES

CONTENTS

Choosing and installing a gas engine. The construction of good gas engines. Examination as to workmanship, running, economy. Reliability and durability of gas engines. Proper erection of a gas engine. Foundation. Arrangement for gas pipes. Rubber bag. Locking devices. Exhaust pipes. Air pipes. Setting up gas engines. Brakes and their use in ascertaining the power of gas engines. Arrangement of a brake test. Distribution of heat in a gas engine. Attendance on gas engines. General remarks. Gas engine oil. Cylinder lubricators. Rules as to starting and stopping a gas engine. The cleaning of a gas engine. General observations and specific examination for defects. The engine refuses to work. Non-starting of the engine. Too much pressure on the gas. Water in the exhaust pot. Difficulty in starting the engine. Irregular running. Loss of power. Weak gas mixtures. Late ignition. Cracks in air inlet. Back firing. Knocking and pounding inside of engine. Dangers and precautionary measure in handling gas engines. Precautions when opening gas valves, removing piston from cylinder, examining with light openings of gas engines. Dangers in starting, cleaning, putting on belts. **Oil Engines.** Gas engines with producer gas. Gasoline and oil engines. Concluding remarks.

120 pages, illustrated, 12mo, cloth, $1.00.

LUBRICANTS,
OILS ❧ AND ❧ GREASES

Treated Theoretically and Giving Practical Informa-
tion Regarding Their

COMPOSITION, USES AND MANUFACTURE

BY ILTYD I. REDWOOD

CONTENTS

8vo, cloth, $1.50.

THEORETICAL AND PRACTICAL

Ammonia Refrigeration

A Work of Reference for Engineers and others Employed in the Management of Ice and Refrigeration Machinery.

By ILTYD I. REDWOOD

CONTENTS

B. T. U. Mechanical Equivalent of a Unit of Heat. Specific Heat. Latent Heat. Theory of Refrigeration. Freezing, by Compressed Air. Ammonia. Characteristics of Ammonia. The Compressor. Stuffing-Boxes. Lubrication. Suction and Discharge Valves. Separator. Condenser-Worm, Receiver. Refrigerator or Brine Tank. Size of Pipe and Area of Cooling Surface. Charging the Plant with Ammonia. Jacket-Water, for Compressor, for Separator. Quantity of Condensing Water Necessary. Loss due to Heating of Condensed Ammonia. Cause of Variation in Excess Pressure. Use of Condensing Pressure in Determining Loss of Ammonia by Leakage. Cooling Directly by Ammonia. Freezing Point of Brine. Making Brine. Specific Heat of Brine. Regulation of Brine Temperature. Indirect Effect of Condensing Water on Brine Temperature. Directions for Determining Refrigerating Efficiency. Equivalent of a Ton of Ice. Compressor Measurement of Ammonia Circulated. Loss of Well-Jacketed Compressors. Loss in Double-Acting Compressors. Distribution of Mercury Wells. Examination of Working Parts. Indicator Diagrams. Ammonia Figures—Effectual Displacement. Volume of Gas. Ammonia Circulated per Twenty-Four Hours. Refrigerating Efficiency. Brine Figures—Gallons Circulated. Pounds Circulated. Degrees Cooled. Total Degrees Extracted. Loss due to Heating of Ammonia Gas. Loss due to Heating of Liquid Ammonia. Calculation of the Maximum Capacity of a Machine. Preparation of Anhydrous Ammonia. Construction of Apparatus, etc., etc.

150 pages, 15 illustrations, cloth, $1.00.

THE SLIDE VALVE

SIMPLY EXPLAINED

By W. J. TENNANT, Asso. M.I.M.E.

REVISED AND MUCH ENLARGED

By J. H. KINEALY, D.E.

CONTENTS OF CHAPTERS:

88 Pages.   *41 Illustrations.*   *12mo, Cloth, $1.00.*

QUICK AND EASY METHODS

OF

CALCULATING

WITH THE SLIDE RULE

A SIMPLE EXPLANATION OF THE THEORY AND USE OF THE SLIDE RULE, LOGARITHMS, ETC.

With numerous examples worked out.

By R. G. BLAINE, M.E.

A most reliable, practical and valuable work for the engineer.

144 Pages.   *Illustrated.*   *12mo, Cloth, $1.00*

The Best and Cheapest in the Market

ALGEBRA SELF-TAUGHT

FOR THE USE OF

Mechanics, Young Engineers and Home Students

BY W. PAGET HIGGS, M.A., D.Sc.

FOURTH EDITION

CONTENTS

Symbols and the signs of operation. The equa
tion and the unknown quantity. Positive and nega
tive quantities. Multiplication, involution, exponents
negative exponents, roots, and the use of exponent
as logarithms. Logarithms. Tables of logarithm
and proportional parts. Transportation of system
of logarithms. Common uses of common logarithms
Compound multiplication and the binomial theorem
Division, fractions and ratio. Rules for division
Rules for fractions. Continued proportion, the serie
and the summation of the series. Examples. Geo
metrical means. Limit of series. Equations. Appen
dix. Index. 104 pages, 12mo, cloth, 60c.

See also **Algebraic Signs,** Spons' Dictionary o
Engineering, No. 2. 40 cts.

See also **Calculus,** Supplement to Spons' Dic